D1250993

TIEBACKS IN FOUNDATION ENGINEERING AND CONSTRUCTION

Harry Schnabel, Jr.

President
Schnabel Foundation Company

McGraw-Hill Book Company

New York St. Louis San Francisco Auckland
Bogotá Hamburg Johannesburg London
Madrid Mexico Montreal New Delhi
Panama Paris São Paulo Singapore
Sydney Tokyo Toronto

Library of Congress Cataloging in Publication Data

Schnabel, Harry, Jr.
Tiebacks in Foundation Engineering and Construction.

Includes index.
1. Anchorage (Structural engineering) 2. Foundations. 3. Retaining-walls. I. Title.
TA772.S36 624.1'5 81-2432 AACR2

ISBN 0-07-055516-8

2 3 4 5 6 7 8 9 0 KPKP 8 9 8 7 6 5 4 3

The editors for this book were Joan Zseleczky and Susan Thomas, the designer was Elliot Epstein, and the production supervisor was Paul A. Malchow. It was set in Trump by Achorn Graphic Services, Inc.

Printed and bound by The Kingsport Press.

CONTENTS

PREFACE

Tiebacks tie a structure to an anchor made in stable soil. The anchor, made deep in the earth at the end of a drilled or driven hole, is the key part of each tieback. In the last two decades, the number of tiebacks used has increased as contractors have perfected sophisticated construction techniques and engineers have developed better understanding of the design. We now understand the technique well enough to organize and present our new knowledge in this first book on tiebacks.

Design and test procedures have evolved over several decades, and have been proven in practice. The book is organized around explaining our procedures, which we have adopted after evaluating and discarding many other possibilities, so that they can be understood and used by structural engineers, geotechnical engineers, students and contractors. The first five chapters are a discussion of the basics: design, construction, testing, and contracting procedures. The next six chapters consider in more detail materials, corrosion protection, test programs, and design of tiebacks to resist uplift, support walls, and stabilize landslides. The proprietary nature of many methods is stressed, and the reader is cautioned that patent infringement must be avoided. A bibliography has been included at the end of the book for the engineer who wishes to read further. This will allow the serious student of tiebacks to evaluate the proposals in this book and consider alternatives not recommended here. Certainly, in the years to come, we expect to improve our techniques, but this book is written based on a conscientious effort to organize and use our present state of knowledge on tiebacks.

I am grateful to many engineers who have contributed to the ideas presented here. Among those from whom I gained particularly significant insights, I want to thank Karl Bauer, Ron Chapman, Jacques Charlier, Ralph Peck, Lymon Reese, Jim Sigourney, Fred Stocker, and Dave Weatherby.

NOTE: In this book, the pronouns *he, him,* and *his* have been used to refer to people in general. This was done for linguistic convenience. It is in no way meant to imply that positions in engineering, design, and construction are filled solely by men.

1

INTRODUCTION

This book is about tiebacks. The use of tiebacks in construction is in its infancy, and, like a child, it will grow rapidly in the next few years. Some of the techniques I describe will quickly become obsolete. Some of the design approaches will be improved. The result will be the greatly expanded use of tiebacks and their application to many areas not now contemplated.

Tiebacks have an anchor in the earth, attached to a tendon, which applies a force to a structure. They are distinguished from deadman anchors by the fact that the anchor is made through the hole drilled or driven to install the tendon. Figure 1-1 shows a sequence of steps used for making one kind of tieback. First a hole is drilled in the ground. A bar is placed in the hole, and the "anchor" end of the hole is filled with concrete. The rest of the hole is backfilled with earth. Wales are installed, and the bar is pulled with hydraulic jacks so that a predetermined force is applied through the wales to the structure. All tiebacks are comprised of the same three basic elements: anchor, tendon, and connection to the structure. The surging growth in their use is due to the development by contractors of methods to construct the tiebacks, and a large part of this book is devoted to describing these methods and explaining how to use them successfully.

Tiebacks have been used frequently to support excavation sheeting walls and tie down slabs. They have also been used for such innovative structures as the cable-supported roof for the Munich Olympic

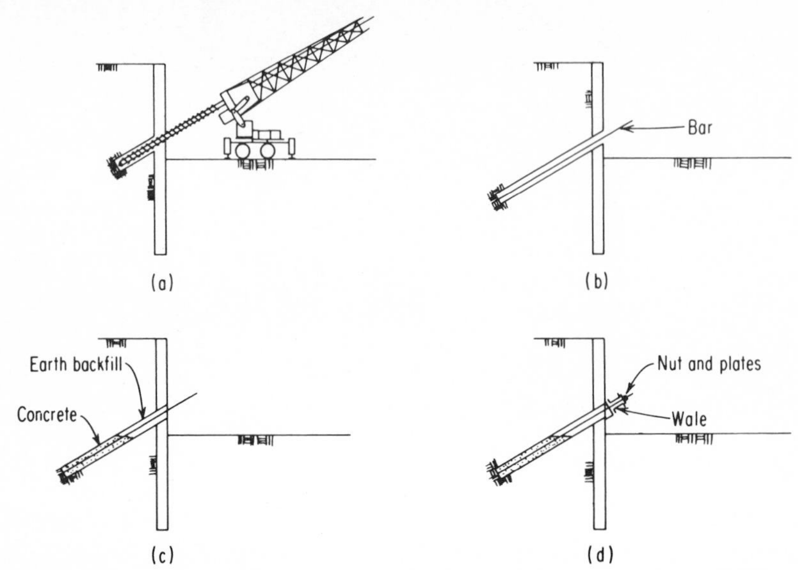

FIG. 1-1 Steps in making a tieback: (*a*) hole drilled; (*b*) bar placed in hole; (*c*) concrete poured for anchor; (*d*) wall connection made.

Stadium and to anchor ski lifts in many parts of the world. They have been protected against corrosion and installed in almost every kind of soil, directly beneath large buildings, and below the groundwater table. These varied conditions have led to the development of a variety of tieback techniques which are designed to meet specific installation and use requirements. We expect these improvements to continue. However no attempt is being made to cover every possibility in this book; instead we seek to cover the basic principles of using tiebacks and the current practice.

Not every reader will want to go into all of the details presented here. The first five chapters have been grouped for the reader who seeks a basic familiarity with the technique and a general understanding of tieback design, construction, and testing.

Presently, the largest use for tiebacks is to furnish horizontal support for excavation bracing walls. The wall shown in Figure 1-2 is a typical application. The only part of a tieback which can be seen is the connection to the wall; the rest is buried in the earth behind the wall. The connection is designed so that the tendon can be overloaded by using jacks to test the tieback and then a permanent force can be locked off in the tendon to support the sheeting wall. There are many connections in use, but they are generally a structural arrangement for

FIG. 1-2 Photograph of a tied-back wall.

distributing the wall loads to the tieback and vice versa, so no attempt will be made in this book to analyze the connection design.

The other end of the tendon is buried in an anchor, which was constructed deep in the ground through a hole made from the face of the sheeting. Many techniques are already in use for making these anchors, and Chapter 3 is devoted to discussing some of those with which I am familiar. Doubtless there are others, and new ones will be developed.

The use of tiebacks to support sheeting walls results in excavations uncluttered by braces and, in some cases, in a faster or less costly project. In general, the more complex the new structure for which the sheeting is being installed, the greater the advantage in using tiebacks to replace braces. For example, braces may interfere with driving piles, drilling caissons, blasting, efficiently setting structural steel, or pouring concrete, especially when preassembled forms are desirable. The project can be done faster in some cases, particularly when tieback installation can avoid the necessity of digging to subgrade to install braces and then removing large quantities of earth beneath long braces. Another instance in which tiebacks save time is when their use allows the job to begin before plans are fully developed by tieing back the sheeting and not having to resolve potential conflicts with the new structure. In some cases tiebacks are less costly than braces. We have found this to be true for many excavations over 30 feet deep where the braces had to react against heels below the subgrade. For temporary sheeting walls, tiebacks have to be cost-effective for the particular project, and their widespread and increased use proves they are.

Another common use for tiebacks is to provide resistance against uplift. One of the first carefully documented uses of tiebacks was to tie down the reservoir side of Cheurfas Dam in North Africa in 1938. This permanent application has been in service for decades. Since then, many more tiebacks have been used to strengthen existing dams. A less dramatic but more common use has been to tie down slabs subject to uplift from hydraulic pressure. We have installed thousands of tiebacks for this purpose alone. Less frequent uses to resist uplift have included the tower legs of an elevated transit system, a television tower, the guy wires from climbing cranes, and the destructive testing of an old elevated structure. In many cases the tieback is more efficient in resisting uplift than additional deadweight concrete, and should be considered as a viable alternative.

There are many techniques used to construct tiebacks. Much of the increased use has resulted from the development of better tieback techniques. In the United States, large drills made it feasible to drill holes 50 feet or longer and 1 foot or more in diameter into the soil. The same drill could bell the deepest part of the hole prior to concreting. The anchor produced would resist pullout by both bearing on the bell and shear on the cylindrical shaft. Another development was the hollow-stem auger, which could be drilled into the ground and then have concrete injected through the hollow stem as it was withdrawn. That would leave a cylindrical shaft in place and allow some pressure to be applied to the concrete as it was injected. Both of these techniques depended for their success on large, expensive equipment which was developed after 1950 for other purposes but which proved adaptable to making some types of tiebacks. In Europe, tieback contractors were developing techniques following a different approach. In general, they used smaller-diameter holes and smaller equipment and injected grout into the ground using high (100- to 500-lb/in^2) grout pressures to form smaller-diameter anchors, which in many cases had greater anchoring capacity. They also experimented with and developed techniques for increasing the anchor capacity using successive injections of high-pressure grout. Another European technique allows a number of bells to be constructed in a small-diameter hole to increase the anchoring capacity. Many of these techniques are protected by patents, and others require specialized know-how or equipment. At the present time there is no "best" way to make a tieback. It depends on the soil, the groundwater, the required capacity, and the capabilities of the engineers and contractors involved.

All of the commonly used techniques result either in a cylindrical anchor, which develops its capacity through shear, or an enlarged anchor, which acts both in shear and bearing. When the cylindrical anchor is made by pouring grout or concrete in an open hole, the shear capacity is limited to the shear strength of the soil or rock. It may be less than that due to contamination of the soil-water contact surface with slick drilling fluids or loose soil, but it is usually limited by the soil or rock properties. In that case, anchor capacity can be increased by drilling a longer anchor hole or, alternatively, increasing the diameter of the hole. It has been common practice in the design of the anchor to assume soil-anchor shear strengths similar to those used in drilled-shaft design. The simplest anchor design is then not much different than the design of a drilled shaft. This type of tieback is not the most widely used though, because the pressure-injected-grout technique results in a cylindrical-type tieback which has a radically increased soil-anchor capacity and is, therefore, less costly. A normal pressure-injected tieback anchor is about 3 inches in diameter by 15 feet long and will resist a pull of 75 tons or more. Obviously proper installation is necessary to get these results, and best results are obtained in granular soils. In any case, where they are suitable, a 3-inch-diameter pressure-injected anchor will have a capacity per lineal foot of anchor three or more times the capacity of a larger 12-inch anchor made without pressure grouting. The rational analysis of the capacity of this type of anchor is still subject to debate, but we cannot escape the obvious fact that it exists. Most are therefore designed empirically.

Some anchors are made with straight shafts and a bell on the end. The capacity of these anchors has been found to be the sum of the shaft shear and bearing on the bell. Normally most of the anchor capacity is developed in bearing, and the design procedure is similar to the procedure for calculating uplift capacity of a drilled pier.

Anchors made with multiple bells are generally designed in still another way. The spacing of bells is fixed so that the bearing of each individual bell is about equal to the shear through a cylinder between it and the next bell. In this way, the anchor capacity is the sum of the bearing of the first bell plus the shear on a soil cylinder whose diameter is the bell diameter and whose length is the distance between the first and last bells.

The design of individual anchors will be discussed in more detail later in the book, but this brief introduction has been included to indicate that the method of making the anchor significantly affects the way in which it acts and, therefore, the way in which it is analyzed. Slight differences in installation procedure or soil encountered can also significantly alter the anchor capacity. For this reason, it is normal to specify a test to an overload on every tieback to verify its anchor capacity before incorporating it in the project.

The tendon is almost always some form of prestressing steel. It is usually connected to the anchor by bond and at the exposed end by standard prestressing hardware. Some infrequently used types of tiebacks are a tendon attached to a plate, point, or tube, which is buried in the anchor and transfers the tendon force to the anchor. In any case, there has to be an effective means of load transfer from the tendon to the anchor.

It is normal for the tieback tendon to have an unbonded or free length. Usually it is desirable to make the anchor deep enough in the ground that soil movements near the supported structure will not affect the tieback capacity. Frequently the plans will specify a block of soil behind the structure in which no part of the anchor can be made. It is then normal to place the prestressing steel in a sheath through this block of soil both to protect it and to prevent it from transferring any load to the particular soil. In some types of tiebacks it is also desirable to keep the anchor in compression, not tension, and the tendon is then protected by a bond breaker through all or a part of the anchor. The tendon thus has a length in which it is not bonded to anything and should be free to stretch or contract elastically. As we shall see later, most of the movement measured during tieback testing is a result of this elastic behavior of the tendon, and the ability to increase and decrease loads systematically gives us a good idea of the actual movement at the anchor. The unbonded length is seen to be important in testing the tieback and analyzing anchor movements.

Threaded prestressing bars and multiple prestressing strands are most commonly used as tendons. These are readily available and have the advantage of large capacity per unit area of steel. They can thus be used in small-diameter holes. They also have high bond values, so they can be anchored by bond into most types of anchors. When that is done, there must be strain compatibility between the tendon and anchor. Often this can only be achieved by minute cracks developing in the anchor. To overcome this necessity of cracking the anchor in tension, the tendons can be secured to the deep end of the anchor and thus create compression in the anchor. For this to occur, the anchorage must be capable of transferring the tendon force to the anchor without cracking or crushing it.

Tiebacks are safe. We test every tieback we install. These tests ensure that tiebacks which will not carry the design load are not used. Our practice is a design and construction type, and we have never had a failure, in use, of a tieback we designed and accepted based on these tests.

The tools to analyze tiebacks are readily available but, more importantly, tiebacks can be tested easily. This then results in a safe application. We will discuss in the next chapter the methods of designing tiebacks to select individual tieback capacities. In this first chapter a brief discussion seems in order as an introduction.

The two chief design requirements are selecting tieback spacing and design load and selecting tieback length so that each anchor is far enough behind the structure. The anchor should not be affected by ground movements related to the intended use. We also check to verify that a block failure cannot occur through the soil which would displace the tiebacks and structure as a unit.

The attainable tieback design load is affected by both the soil and the techniques used. Often the best course is to select the total or line load required and let the tieback installer select the method and design load per tieback. We have had simple projects where three different types of tiebacks were used, just as a result of soil variations. Obviously in such a situation, if a method specification had been used, most of the tiebacks would have failed or the job would have cost a lot more, because the tiebacks' design loads would have had to be determined by the most unfavorable condition. However, if the contractor is really going to design the tiebacks, and that is how it is usually done, then the designer of the structure must have good tools to check the tieback capacity and length. Fortunately, such tools are available.

We have tried to develop a standardized test for our tiebacks and feel we have been successful. Chapter 4 discusses in detail the test procedure we use, compares it with some others, and discusses in more detail the significance of the results we obtain. In general, a center-hole hydraulic jack can be used to apply a force to the tieback. We test most tiebacks by increasing that force in steps and simultaneously measuring the movement of the tieback. For our routine proof test the tieback is tested to a maximum load equal to 120 percent of the design load. After testing is completed, the jack force is reduced to some lower amount, and the tendon secured to the structure in a way that maintains that force on the tieback. Since prestressing steel is used at high stress levels, and consequently has a high elastic strain to reach its working load, the lock-off load is usually our best estimate of the actual load in service. To prevent excessive movements, we have to preload the tieback anyway, and this simple test procedure involves little extra work but gives a valuable load/elongation curve for every tieback. We run a more detailed performance test on the first few tiebacks on each project. This test is to a higher overload, usually 133 percent of design, and involves cyclically loading and unloading the tieback through each of the steps of the first test, plus the higher overload. From this we can get a load/elongation pattern, estimate with reasonable accuracy the pattern of anchor movements, and determine that our tendon is substantially free to strain in the unbonded length. The load/elongation patterns of proof and performance tests should be substantially the same, and they can be used together to interpret the test results. If we are working with a familiar type of tieback in an appropriate soil, these tests should let us analyze the performance of all the tiebacks.

Center-hole load cells make long-term checking of tieback loads very easy. They can easily be placed under the test jack and have the lock-off hardware bear against them. The load cells are relatively inexpensive and have good long-term characteristics. They are also easy to read. They are concentric with the force; thus it is much easier to measure brace loads with load cells than with instrumentation. With load cells we can easily measure the tieback load in use.

Optical surveys are usually precise enough to check movement over a long period of time. On sheeting jobs it is usually sufficient to have a mark on the tieback. Similar marks on other types of structures are usually adequate.

Where greater precision on movement readings is desired, an extensometer is often the best tool. This is simply a device anchored in the soil far enough beyond the ends of the tiebacks that we can consider it fixed. The simplest form is a rod in a sleeve so that the rod is unaffected by soil movements near the structure, and can thus be used as a reference point. With such a setup, movements can be read on micrometers or dial gauges to within a thousandth of an inch. More sophisticated extensometers can be used to read movements at several points in the soil or even at locations in the anchor. As a field control device they are not often required.

Instrumentation is readily available to measure movement of the tiebacks and supported structure. Other instruments easily measure the load in the tieback at any time. The result is that tieback jobs are easier to instrument than most other systems with which they compete, and they are routinely more thoroughly tested. From this process of design, construction, and careful testing has come an iterative approach which has led to prompt feedback of the test results into the design and construction process.

One of the earliest results observed was that movements were often less than with braces. Obviously if the tieback is preloaded to its most probable working load, that force is applied to the structure. No movement of the structure is required to develop that force. In many cases no movement of the structure can occur until that force is exceeded. As a result observed movements are often far less with tiebacks than with braces. This is not to claim preloading eliminates movement. But it apparently reduces it substantially.

When using tiebacks, we have a tool that has a history of successful use and can be individually tested so that every tieback, not just a sampling, is capable of carrying the design load, can be analyzed in conjunction with the structure, and can easily be observed for long-term reliability. We also have a tool with some limitations which should be considered.

One of the restrictions on the use of tiebacks is that they are often not placed on the property of the owner causing the work to be done. Since excavation sheeting is often placed near the property line, the tiebacks are often on a neighbor's property. Obviously permission must be obtained for placing the tiebacks, and they can only be used if it is granted. In many metropolitan areas this is not a major problem because the tiebacks are placed beneath adjacent streets and the cities routinely grant permission.

Another problem occurs when there is a structure that interferes with tieback placement. For example, there may be a subway structure in the street that is too close to allow the tiebacks to develop sufficient length to be stable. A building across a narrow alley, a maze of utilities under a street, and many other physical problems can prevent them from being used effectively.

The soils may not be suitable for economical tiebacks. For example, at the present time soft clays are the most troublesome for tieback construction. Tiebacks in such soft soils will be much costlier per ton of capacity than in better soils, and as a result the installed cost may compare unfavorably with alternative ways of performing the project.

When tiebacks must be installed under adjacent structures, it may be prudent to select a method where no ground can be lost. Thus, there would be no chance of settling of the structure from tieback construction. Such methods are available, but are usually proprietary.

Soil stratification may also affect the type of tieback chosen. It has been our experience that stratification can have a more serious effect on some types of tiebacks than others. For example, a drilled and belled tieback is made with an open hole. If the hole will not stand open, that sort of tieback cannot be made. Even if only a few feet of the hole is in a caving stratum, other tiebacks may be more effective in the stratified soil.

Groundwater, both quantity and elevation, may seriously affect the selection of tieback type. There are a number of techniques for installing tiebacks in soil below the water table, and one of these would have to be used.

Noise might also affect the method selected. Some methods are noisier than others, and in some situations noise during installation could be the controlling factor.

Corrosion, or the need to resist corrosion, can seriously affect the selection of tieback type. Tiebacks have a relatively short history of use as part of a permanent structure. So from experience we cannot know too much about their long-term resistance to corrosion. What we now do is analyze the potential for corrosion, design to protect against it, and use a tieback type that will allow the use of the proper corrosion protection. A large number of judgments are involved, and as a result I am cautious in designing projects incorporating tiebacks for long-term use. For example, we have not used tiebacks to support bulkheads in a seawater environment. Because of our caution, we have never had a corrosion failure of a tieback we designed. We have also on several occasions installed tiebacks in a corrosive environment where we carefully followed someone else's method specification, and to date they seem to be performing well. However, we feel it is too early to decide the long-term life of the tiebacks, and we are still cautious when it comes to designing tiebacks for long-term use.

Most of the tiebacks we have installed are anchored in granular or preconsolidated clay soils or in rock. We have not as yet had a problem with creep, slow movement of the anchor. The methods for dealing with creep involve an overload, an estimate of long-term creep, and a lock-off load high enough that anchor creep will not have serious effects on the structure or reduce the tieback load below the appropriate lock-off. When these estimates are based on field testing, they require a precision that is even difficult to obtain in the laboratory. So if creep is felt to be a serious problem, it may preclude the use of tiebacks. The alternative is to use a creep test of great complexity.

From the last few paragraphs it should be clear that, although the use of tiebacks is increasing rapidly, it is not always appropriate. One purpose of this book is to help clarify for the potential user where the proper place to use tiebacks is and where it would not be wise to attempt them. Next we will discuss tieback design, construction techniques, and test programs and give a brief summary for the user, who will then be able to make intelligent decisions about their use. The rest of the book will cover tieback materials, corrosion protection, testing, and common applications in more detail.

Personal recognition of my limitations, combined with the present state of the art, have caused me to write this book in the first person. It is not intended to be a complete, authoritative work as much as it is a personal observation of the early stages in the growth in the use of a new and promising construction technique. As such, opinions may be expressed which are based solely on personal feelings. It is, however, written with the hope that it will help engineers develop a familiarity with tiebacks that may encourage them to permit their use in future structures. It may also stimulate others to help answer the many remaining questions. With that in mind, let us proceed to a review of design.

2

DESIGN

Better tiebacks have led to increased use. Most of the improvements have come from specialist organizations that design and install the tiebacks. Today, these more sophisticated tieback techniques dominate the market. A competent engineer must know when to incorporate these techniques in his design. The principles of design must be understood to evaluate any tieback, and this chapter addresses the important tieback design principles.

We use the terms tendon and anchor for the two most important elements of the tieback. Figure 2-1 shows these basic tieback elements. A hole is made from the structure by driving or drilling. The anchor is made in this hole in such a way that it anchors the tendon to the soil or rock. The other end of the tendon is secured to the structure, in this case a wall. The purpose of the tieback is for the tendon to apply a force to the structure, and this force is usually applied by stressing the tendon with jacks. The tendon must not only have adequate capacity in tension, it must also be secured to the anchor and structure with at least that capacity. The anchor must be designed so it will not pull out of the soil. It must be located so the soil-structure mass it reinforces is stable. The factor of safety of the tendon and tieback should be compatible and adequate. Finally, every tieback should be stressed and tested, and, since design is often a choice between available systems, a proper testing procedure is often the specification requirement which is most critical. These four portions of tieback design—anchor, tendon, factor of safety, and testing procedure—will be discussed.

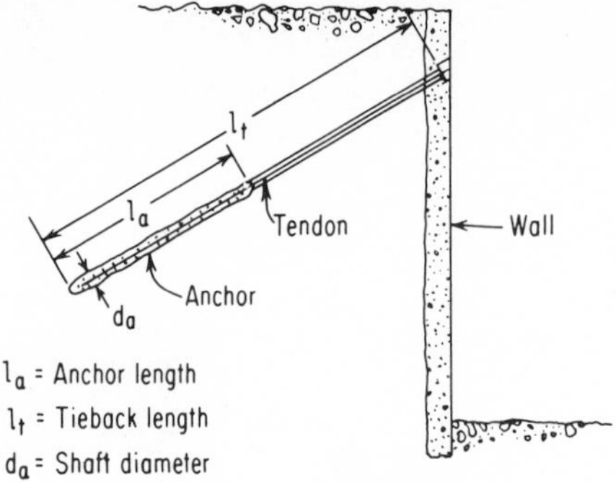

FIG. 2-1 Shaft tieback.

Anchors

Anchors generally can be divided into shaft type and underreamed or enlarged types, based on their shape. They can also be divided into anchors for which the grout or concrete was pumped using large pressures to improve the soil properties and anchors for which little or no pressure was used. This results in four basic types of anchors which we must consider.

The earliest tiebacks we made were in soil or rock which was easily drilled and in which the hole would stand open without casing or other support. These anchors were made by pouring concrete in a cylindrical hole, which resulted in a shaft-type anchor made without grout pressure. The anchor shown in Figure 2-1 is a typical shaft-type anchor whose capacity is calculated using formula 2-1.

$$P_s = A_c \times f_g = \pi \, d_a l_a f_g \qquad [2\text{-}1]$$

where P_s = shaft-anchor capacity
A_c = the cylindrical area of soil-anchor bond
f_g = the unit soil-anchor bond stress
d_a = diameter of the shaft anchor
l_a = length of the shaft anchor

The diameter and length of the anchor are determined from the size of the grout or concrete body. An average value of bond stress between the soil and anchor is used. If the shear stress in the soil just outside the anchor is less, of course this should be used. Our experience is that drilling techniques smear or soften the surface, so the bond stress is often less than the shear strength. We also know that the actual bond stress is a function of anchor strain. Curves *A* and *B* on Figure 2-2 represent two common conditions. Curve *A* represents a soil where

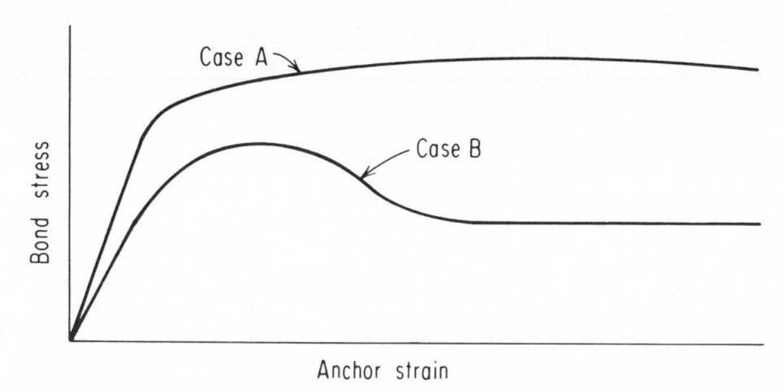

FIG. 2-2 Two typical cases of stress-strain behavior of an earth anchor.

very little strain is necessary to mobilize most of the soil-anchor bond stress and where the stress continues to increase or remain stable with increased movement. Curve *B* is a softer soil which never develops as much bond stress, requires more strain to reach its maximum, and softens as that strain is exceeded so that its final bond stress is less than its maximum. The unit soil-anchor bond stress we use in this formula is the average which will act on the anchor surface.

Experience with tiebacks suggests that the design soil-anchor bond stress will be about 1000 lb/ft^2, except in softer clays. In these softer soils we use the values tabulated in Table 2-1. If larger values can safely be relied upon, this can be established in the field with the testing program. Tiebacks can also be constructed so the anchor is either in tension or compression, and this may affect the unit soil-anchor bond stress. In sandy soil, the anchor capacity does not seem to be affected by the overburden stress. The anchor body may also be in several strata with different properties. For these reasons and others, we believe this formula should only be used to select a reasonable anchor size. Anchors of this size are then verified by a field testing program. If

TABLE 2-1 Typical Values for Unit Soil-Anchor Bond Stress

Soil type	Standard penetration resistances (blows/ft)	Bond stress between anchor and soil (kip/ft^2)
Silty clay	3–6	0.50–1.00
Sandy clay	3–6	0.75–1.00
Medium clay	4–8	0.75–1.25
Firm clay or stiffer	over 8	1.00–1.50

TABLE 2-2 Typical Values for Unit Rock-Anchor Bond Stress

Rock type	Bond stress between grout and rock (lb/in²)
Sandstone	120–250
Soft shales	30–120
Slates and hard shales	120–200
Soft limestone	150–220
Hard limestone	300–400
Granite and basalt	250–800

modifications to the anchor are necessary after analyzing the field tests, usually the capacity can be increased by increasing the length or diameter.

Rock anchors are usually shaft anchors made with no pressure on the grout. This is probably because the anchor capacity is developed so easily unless the rock is weathered, highly jointed, or full of solution passages. Short holes drilled with ordinary rock drills develop the tendon capacity. Table 2-2 lists representative values for the bond stress. These values are about 10 to 100 times those listed in Table 2-1. For example, we could expect an anchor made in a 4-inch drill hole to have a capacity per lineal foot in sandstone of

$$P = \pi\, d_a l_a f_g = \begin{cases} \pi \times 4 \times 12 \times 120 = 18{,}096 \text{ lb/ft minimum} \\ \pi \times 4 \times 12 \times 250 = 37{,}700 \text{ lb/ft maximum} \end{cases}$$

Rock anchors can also be made with epoxy grouts, which have even higher bond stresses in most rock. However, hole-drilling problems restrict their use in most cases.

The second type of anchor, underreamed and poured without pressure, can only be used where the hole will stand open. Figure 2-3 illustrates a typical underreamed, or belled, tieback. The shaft is usually drilled to the full depth, then a special tool used, which enlarges the hole by scraping soil off the sides for the bell. The tendon is placed, with a plate at its end, and the concrete poured or pumped into the bell. Usually, we concrete a portion of the shaft. The capacity of the anchor comes from both bearing on the area of the bell and shear on the shaft. Where the shaft length is short, d or less, it is usually ignored. The bearing capacity of a belled anchor is:

$$P_b = A_b p_b + A_s f_g = \frac{\pi}{4}(d^2 - d_a^{2}) N_c C + \pi\, d_a l_s f_g \qquad [2\text{-}2]$$

where P_b = bell anchor capacity
A_b = bearing area of the bell
p_b = the bearing pressure at the bell
d = bell diameter
d_a = diameter of the shaft
N_c = bearing capacity factor (9 for a deep buried footing)
C = cohesion of the soil at the bell
l_s = length of anchor shaft
f_g = unit soil-anchor bond stress

The bell is usually plain concrete. The concrete of the bell is extremely difficult to analyze and, in practice, is usually not designed but selected from experience.

A belled tieback should be checked for the shear capacity of the soil along a cylindrical surface through the edge of the bell. The length of this cylinder must be long enough to develop a total shear which will equal or exceed the anchor capacity in bearing.

These first two types of tieback anchors can only be used in favorable conditions. Each type requires a hole which can be easily drilled and will stand open until the tendon and concrete are placed. The maximum anchor capacity is limited by the in-situ soil strength. For these reasons, they represent only a small fraction of the earth tiebacks in use today.

Most modern tiebacks use pressure-injected grout to increase the anchor capacity by improving the soil. They also utilize a variety of driving or drilling techniques, so they can be used in all but the softest soils. Chapter 3 will discuss some of the construction techniques in more detail. Most of these techniques combine specialized drilling and grouting procedures, require special equipment or materials, and have

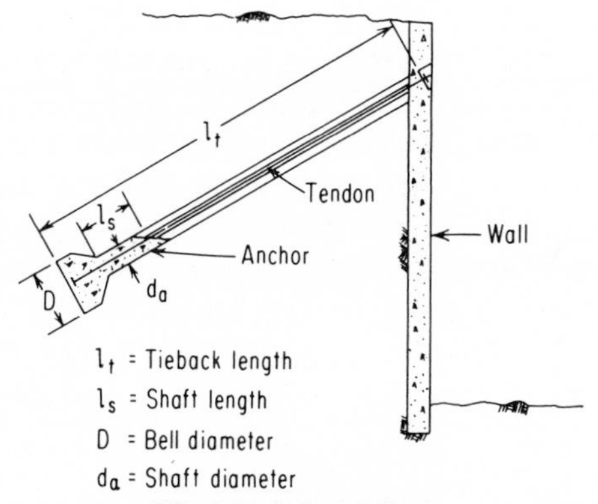

FIG. 2-3 Belled tieback.

been developed as proprietary products of organizations which specialize in making tiebacks. Various organizations have developed competing systems. Each has developed methods for evaluating the anchor capacity it can make in a particular case. The tiebacks made with these techniques are dependable and economical. As a result, throughout the world most tiebacks are proprietary products installed by specialists.

The first tiebacks using grout pumped under high pressures were made in Germany by Karl Bauer in 1958. Dr. Bauer was granted a patent on his technique, which has since spread around the world. The essence of the technique is:

1. Drive a pipe with a knock-off point into sandy soil.
2. Place a bar tendon in the pipe and knock off the point.
3. Simultaneously withdraw the pipe for a portion of its length while pumping grout through it and maintaining grout pressures of several hundred pounds per square inch against the soil.
4. Stop pumping grout and withdraw the rest of the pipe.

The grout is a neat cement grout. The high pressure can be maintained because the soil around the driven pipe is an effective seal. The anchor made with this technique usually has a diameter slightly larger than the drive pipe. The anchor length is the length of soil against which high-pressure grout was pumped while the pipe was withdrawn. A typical anchor with this technique is slightly more than 3 inches in diameter and 15 feet long. Figure 2-4 is taken from data based on tests of thousands of anchors, reported by H. Ostermayer at the 1974 London Conference on Diaphragm Walls and Anchorages. It is clear from Figure 2-4 that a small anchor made with this technique has a large capacity. We have found charts such as this the most helpful way of selecting anchor capacity. Anchor-soil bond stresses would often exceed 10,000 or 20,000 lb/ft^2, if formula 2-1 is back-figured. Clearly, the in-situ soils listed are not capable of developing a shear strength of this magnitude. The pressure of the grout on the soil has dramatically improved the capacity of this type of anchor.

I believe this improvement results directly from use of the technique developed by Dr. Bauer. As the grout is pumped, it exerts a radial pressure on the soil. The sandy soil has been densified by driving the pipe through it, and is particularly dense just next to the pipe. Even at the high pressures used, the grout will not penetrate the sand more than a tiny fraction of an inch before the sand blocks the passage of the

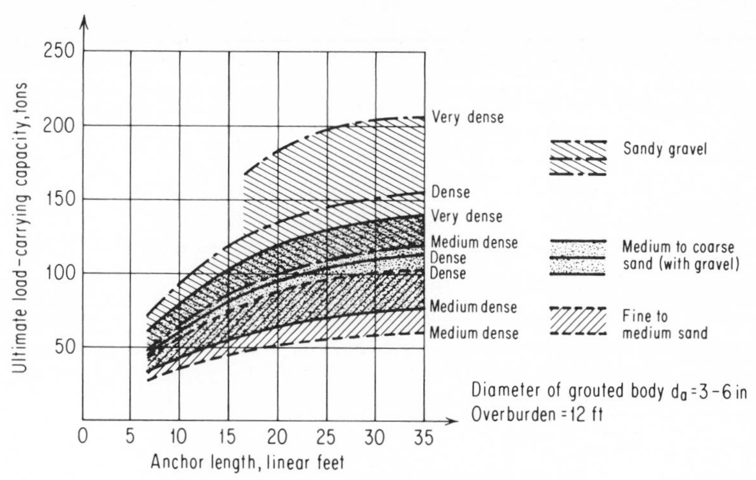

FIG. 2-4 Capacity of pressure-grouted anchors.

cement particles. The water in the grout is separated from the cement and passes into the sand. This leaves a film of cement on the sand surface. As the process continues, a filter cake of cement is built on the sand surface through which water continues to pass but which retains all of the cement. The pressure of the grout acts on this filter cake, and the sand must act with an equal and opposite radial confining pressure since it is not displaced. The filter cake increases in thickness until it finally occupies all of the space between the bar and the sand. This mass of cement has become a solid by driving off water, but retains enough water for hydration. The resulting grout anchor is extremely hard and dense, but more importantly it is still being acted on by the radial stresses in the soil. These high induced stresses dramatically improve the strength of the soil and the soil-anchor bond stresses when the tendon is subsequently loaded. For example, if $\tan \phi = 0.5$ and the induced radial stress is 300 lb/in^2, a soil-anchor bond stress of 150 lb/in^2 or 21,600 lb/ft^2 would be reasonable. The mechanism which produces this increased capacity depends on porous soils, such as sand and gravel.

We use at least two other methods of making shaft-type anchors in which we pump grout under pressure. In one system, a hole is drilled into the sand and filled with fluid grout. An anchor assembly is slid into the hole, where it displaces some of the grout. Grout is first pumped into a porous bag which surrounds the tendon at the structure end of the anchor length. This bag expands against the soil and allows the water, but not the cement, to pass through it. After a short time, it becomes filled with cement particles which are pressing against the soil, much as in the Bauer technique. The purpose of this first grouting is to seal the anchor portion so high-pressure grout can be pumped into it. Visualize the bag full of grout as performing the same function as the pipe in the Bauer technique. Then grout under high pressures is pumped into the anchor length. The anchor performs in much the same way as the first case, with bond stresses an order of magnitude greater than would be expected from the in-situ properties.

Less dramatic benefits are obtained by pumping grout with pressure when making hollow-stem-auger tiebacks. Many drilling techniques, including the auger, tend to smear the sides of the hole with a softer soil. The hollow-stem auger allows grout to be pumped through the auger as it is withdrawn. It has been our experience that this generally overcomes the smear problem and allows the anchor-soil bond stress to be based on the soil shear strength. Anchor diameter is fixed by the auger, and the length is then selected to give the desired anchor capacity.

The fourth general type of anchor is made by injecting grout at high pressures to form bulbs or enlargements in the anchor. The high-pressure grout will displace cohesive soils; so large quantities of it can be pumped. Many tiebacks are made by controlling the pressure and quantities so that a desirable shape can be obtained. In this way, anchor capacity can be increased. Some benefit is also obtained by increasing the radial stresses, but in a clay these will largely be neutral stresses carried by the water. These anchors are generally referred to as multiple-injection anchors.

They are made by drilling a hole in the ground, filling it with grout, and later injecting additional quantities of grout in the anchor zone. The resulting anchor has increased capacity, but the exact mechanism is not well understood. The improvement can be visualized as the result of either increasing the effective diameter of the anchor by forming a series of bulges or strengthening the clay around the anchor by subjecting it to high pressures, which are first carried as neutral pressures but lead to high effective stresses as the water migrates radially away from the anchor. By repeating this reinjection process for several days with small quantities of grout pumped under high pressure, the clay strength around the anchor is increased. The increase for this type of anchor is not of the same order of magnitude as for an anchor in porous soil, but it is significant. At this time, the only way we can estimate the capacity of this anchor is from experience. As a result, several anchors are usually made at the site and tested to failure to establish capacity. By increasing the number of successive injections, we can increase the size of the bulb or the effective radial stress and may be able to increase the tieback capacity.

All of the techniques for multiple injection of grout to form anchors with increased capacities were developed by particular organizations. Much of the increase in the number of anchors used has resulted from the success of this type of anchor. Often, the only people who have sufficient understanding of the particular technique to make it work and the specialized tools required work for these organizations. They can evaluate the advantages and disadvantages of their particular systems. They can give reasonable estimates of anchor capacities for a particular job. I feel it is only prudent to consult them.

Anchor location is as important as anchor capacity. The anchor should be made beyond a minimum distance from the structure it supports. It should be long enough that it is anchored in soil which will not fail from the loads applied by the structure. The anchor must also be in a soil appropriate for the construction technique. In the design of tiebacks, all three of these conditions must be considered.

The distance between the anchor and structure is an important part of the design. This is readily seen in Figure 2-5. For our tiebacks supporting excavation bracing, we do not want to develop any of our anchor capacity in the soil between the critical failure surface and the wall. If we did, movement of the wall might reduce anchor capacity, which would lead to more wall movement and ultimately collapse. From this example it can be seen that, if the tieback is to exert a dependable force on the structure, there may be a requisite minimum distance between the anchor and the structure.

There is no universal agreement on placing the anchor beyond a minimum distance from the structure. Some companies construct the anchor in the example above right to the back of the wall. This creates a tensile reinforcement of the soil in the anchor zone. Soil nailing and other techniques are also based on this general type of analysis. At this time, not enough data is available to prove that this is a better way to

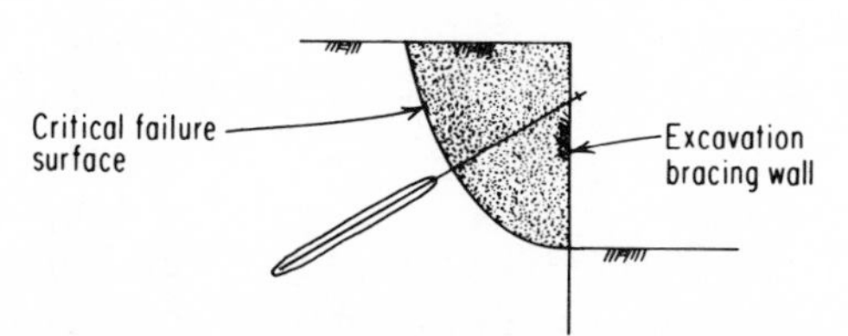

FIG. 2-5 Example of soil in which anchor should not be placed. Anchor should not be made in soil represented by the shaded area between the critical failure surface and the wall.

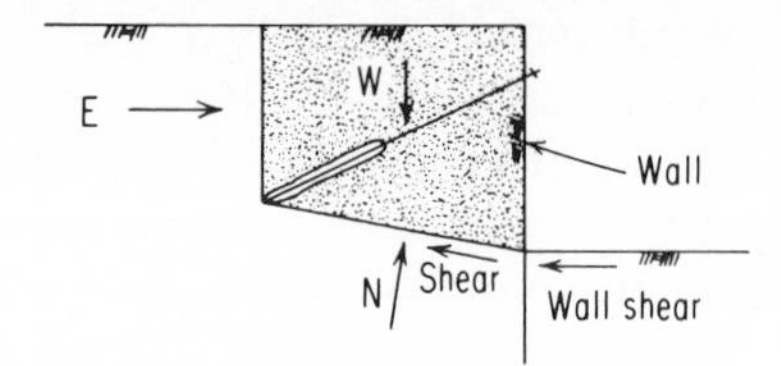

FIG. 2-6 Free body for mass stability.

model the soil behaviors; so we continue to design our anchors to be beyond the active zone.

The total length of the tieback is important. As in Figure 2-6, the mass of soil which is reinforced by the tieback can be visualized as a free body which must be stable. The external forces which act on the free body are often determined by the soil properties. The design principle is to locate the anchor so that this free body is always stable. For the general case in Figure 2-6 we have determined that the failure will occur through the end of the anchor. The shape of the soil mass should be determined from the soil properties. It is not necessary to do this precisely, because large increases in the calculated values can be made relatively easily by changing the length of the tieback or the angle at the structure. The general stability equations, $\Sigma H = 0$, $\Sigma V = 0$, and $\Sigma M = 0$, have to be satisfied. To prevent sliding, the sum of the horizontal forces resisting sliding must exceed the sum of the forces causing sliding by the desired factor of safety. To prevent overturning, the sum of the moments must be similarly checked. For the case of uniform sand and a spiral failure surface, this problem has been solved exactly. The solution is complex and time-consuming and depends on soil conditions which we have never encountered on a project. We find the best way to solve for mass stability is to analyze a free body on the basis of our understanding of the soil behavior and the most probable anchor location. If this mass does not have an adequate factor of safety, we change its shape by altering the tieback location. This is simply a matter of increasing the length of the tieback or changing the angle at the structure. Then we redetermine the boundaries of the mass and recheck. This is not usually a difficult condition to satisfy, but is a design requirement that is often overlooked. Properly done, it assures us that the mass of soil tied together by the tieback is stable.

The anchor must be located in a soil where it will develop the required capacity. This is the third important consideration of anchor location. Frequently the tieback angle or length is modified in the field to cause the anchor to be made in the proper soil. In most cases, this is done by the contractor on the basis of observations made during installation. Some flexibility has to be provided here, since soils at most sites vary. If the designer addresses this problem, it is usually by establishing allowable angular deviation at the structure and a maximum allowable length. These should be generous allowances, or they will be of little value.

These are the main considerations on the design of the anchor. Of course, in every case a tendon must be secured to the anchor. The tendon is steel, stretched between the anchor and structure, and it applies the same force to both structure and anchor. Next, let's consider the tendon.

Tendons

Modern tiebacks all use some form of prestressing steel for the tendon. The threaded bar and seven-wire strand are most often used. Design of the tendon involves selecting steel with proper tensile capacity, connecting it to the structure and the anchor so that it can be preloaded, preventing it from bonding over a part of its length nearest the structure, and providing for it to apply a force on the structure, which is often not colinear with the applied force it is designed to resist. Adequate corrosion resistance for the life of the project must also be included. Figure 2-7 is a close-up photograph of a bar and also of a typical strand tendon made of several strands.

Both types of tendon shown in Figure 2-7 use readily available steel designed for prestressing. These steels are designed to be used at high stresses for long periods of time. Their metallurgy has been selected for the best combination of high capacity and corrosion resistance, but, since these may be mutually exclusive criteria, neither may be fully satisfied. A later section of this book discusses ways to coat the bars to prevent them from corroding. These bars and strands are classified according to their guaranteed strength. The design and preload are fixed as 60 and 80 percent respectively, when used for prestressed concrete. For years we have used these same criteria for design capacity and maximum test load with excellent results. The bars are made in a family of sizes. Table 2-3 is a schedule of tendon design loads based on $f = 0.6\ f_{pu}$, where f is the allowable tendon design load and f_{pu} is the tendon guaranteed strength. When using seven-wire strands, we use Table 2-4 to select the proper number of strands. We have found in the last few years that these tables include every tendon we have used.

Selection of tendon capacity is simply choosing a tendon from these

FIG. 2-7a Strand tendon.

FIG. 2-7b Bar tendon.

tables which is compatible with the technique for making the tiebacks. Familiarity with the tieback procedure is essential. For example, we prefer to use the bars in our pressure-injected anchors.

If a test load in excess of 133 percent is specified, the tables cannot be directly used to select the design load. A correction, based on formula 2-3, must be applied to the design load shown.

$$\text{D.L.} = \frac{1.33}{\text{T.L.}} (\text{D.L.}_T) \qquad [2\text{-}3]$$

where D.L. = the reduced design load

D.L._T = the design load shown in Table 2-3 or Table 2-4

T.L. = the ratio of the specified test to design load, but not less than 1.33

This formula is based on the overload during testing never exceeding

TABLE 2-3 Common Bar-Tendon Capacities (kips)

Diameter (inches)	Ultimate strength (kip/in²)	Guaranteed strength f_{pu}	Maximum test load $0.80f_{pu}$	Design capacity $0.60f_{pu}$	Cross-section area (in²)	Weight (lbs/ft)
1	150	127.8	102.2	76.7	0.85	2.96
1	160	136.3	109.1	81.8	0.85	2.96
1¼	150	187.5	150.0	112.5	1.25	4.40
1¼	160	200.0	160.0	120.0	1.25	4.40
1⅜	150	234.0	187.2	140.4	1.56	5.31

TABLE 2-4 Tendon Capacity Calculator for ½-inch Seven-Wire Strands (kips)

Number of strands	Guaranteed strength f_{pu}	Maximum test load $0.80f_{pu}$	Design capacity $0.60f_{pu}$	Cross-section area (in^2)	Weight (lbs/ft)
1	41.3	33.0	24.8	0.153	0.525
2	82.6	66.0	49.6	0.306	1.050
3	123.9	99.1	74.3	0.459	1.575
4	165.2	132.2	99.1	0.612	2.100
5	206.5	165.2	123.9	0.765	2.625
6	247.8	198.2	148.7	0.918	3.150
7	289.1	231.3	173.5	1.071	3.675
8	330.4	264.3	198.2	1.224	4.200
9	371.7	297.4	223.0	1.377	4.725
10	413.0	330.4	247.8	1.530	5.250
11	454.3	363.4	272.6	1.683	5.775
12	495.6	396.5	297.4	1.836	6.300
13	536.9	429.5	322.1	1.989	6.825
14	578.2	462.6	346.9	2.142	7.350
15	619.5	495.6	371.7	2.295	7.875

80 percent of the guaranteed steel strength. Occasionally tests up to 90 percent are permitted, but this must be done with great care.

In most of the tiebacks we use, the tendon is secured to the anchor by bond. Both the deformed bars and the strand have surfaces which are excellent for bonding. We have tested both and find that only a few feet of bond will develop the full tensile strength of the tendon. In some cases, it appears that some of the strand at the front of an anchor will debond because of the greater strain, but this has caused no problems with its use. As a precaution, the strand is often used with spiders which cause it to be alternately spread and bundled, as shown in Figure 2-8.

Mechanical anchorage of the tendon may be desirable. We use a plate on the end of our tendons in belled anchors. A mechanical anchorage is also used on "compression-tube" anchors. These are multiple-injection anchors specially designed to keep the anchor grout

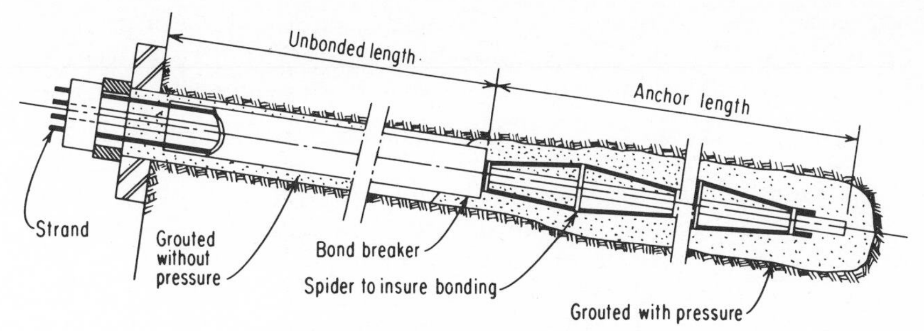

FIG. 2-8 Tieback with strand tendon.

body in compression. This is done by securing the end of the tendon to a specially machined pipe which is designed to be embedded at the end of the anchor. The pipe is designed to transmit the tieback design load to the anchor by a combination of bond and bearing.

At the other end, the tendon is secured to the structure with standard hardware. We try to use hardware furnished by the tendon supplier and not design it. We also design the structure to accept the tieback and test loads. Such structural design will not be covered here, except to state that the function of the tieback is to apply a force on the structure. The most efficient use of the tieback is to have it as nearly colinear with the required force as possible. This is the end of the tendon where the tieback test must be performed. It must be accessible to workmen for testing and preloading.

A portion of the tendon may be prevented from bonding to anything. This is usually accomplished by covering it with a plastic pipe. Then whether the pipe is buried in soil, weak grout, or even the anchor, the covered steel is free to stretch elastically. We call this length of tendon the unbonded length. The unbonded length is an important part of the tendon design. The steel in this length should behave elastically.

The design of the unbonded length is often a function of the tieback technique. Compression-tube anchors use a tendon which is completely unbonded. For this type of anchor, the tendon is mechanically connected to the compression tube at the end of the tieback. It is covered with a pipe for the rest of its length, so that the entire tieback force can be transferred to the anchor by the compression tube. For our belled anchor, we use a bearing plate at the end of the tendon, but usually allow the tendon to be bonded to the concrete of the bell. We accomplish this by covering the tendon with a pipe between the structure and the near end of the bell. For our hollow-stem-auger tiebacks, we may extend the pipe some distance into the anchor. Since tendon bond to the front of the anchor is prevented, this part of the anchor goes into compression when the tendon is stressed. Cracking or discing of many anchors is a result of tensile-strain incompatibility between the anchor body and tendon. This can be avoided or reduced by continuing the unbonded length of the tendon into the anchor, which puts the near end of the anchor in compression.

For some tiebacks, it may be desirable to have a full-length anchor and no unbonded length. As one example, the reinforcing provided in a landslide area may be desirable for stabilizing the soil mass.

Most tiebacks have a tendon with some unbonded length. The tieback should be preloaded to strain the tendon in the unbonded length as well as the anchor. If the tieback is not preloaded, the structure must move the same amount to load the tieback. Since the elastic strain in the unbonded length is frequently of the same magnitude as the allowable structure movement, preloading reduces these movements.

Factor of Safety

Tendon and anchor capacities must have factors of safety. Since each tieback must be preloaded, we perform a simple effective test on it. Tiebacks are the only construction technique we use where each item is individually tested. When considering the factor of safety, we look both at the probability of the design load being exceeded and the possible amount. Then we consider the amount the tieback ultimate capacity has to exceed the design load to prevent failure. Whenever possible, we use the steel at the same working stresses allowed in the American Institute of Steel Construction (AISC) code or the Prestressed Concrete Institute (PCI) code.

The tendons are usually designed for a factor of safety of 1.67. This is the normal overload used in steel design. The mechanical connections are also designed with at least that factor of safety. These are manufactured products with a rigorous inspection system. We have found the actual strength always exceeds the guaranteed strength. There is very little uncertainty in the factor of safety of the steel.

The anchor should be designed for a factor of safety compatible with that of the tendon. The tendon, which is a product manufactured with rigid controls, usually has a factor of safety of 1.67. This should be the lowest factor of safety considered for the anchor. Because we cannot be as sure of the processes used for making the anchor and soil is a much more variable material than steel, a higher factor of safety may be used for designing the anchor. In any case, the ultimate strength of the anchor should be designed to equal or exceed the ultimate strength of the tendon.

All of this discussion is based on the assumption that no factor of safety is included in the design loads, but that the design loads are also readily established. In other words, the design loads are calculated from the maximum force from the structure which the tiebacks must carry. If these can be calculated with the same exactness as the loads which will normally come on a steel or concrete building, we use them as our design load for selecting the tiebacks. If there is a greater degree of uncertainty, then we may have to increase our design load to account for it or increase the factor of safety of the tiebacks. We only increase the design load or derate the tiebacks on the basis of a careful study.

For most tiebacks we install, we design the steel for a maximum working stress that is 60 percent of the yield (or guaranteed strength) and provide an anchor with a matching factor of safety.

Testing

We then test each tieback to verify it will carry an overload and preload it. The test procedure will be discussed in more detail in Chapter 4. The preloading has an important impact on design. It is very simple to do and should normally be done. We use a jack, acting on the structure, to pull on the tendon. Then when the predetermined preload is recorded on the jack, we secure the tendon to the structure. When the jack is removed, the tendon remains stressed and applies equal and opposite forces to the structure and the anchor. We use a test procedure which employs two tests. Most tiebacks are proof-tested. While the tieback load is increased in increments, movement of the end of the tendon is measured. The maximum test load is usually 120 percent of design. Then the load is decreased in decrements until the desired tieback preload is reached, after which the tendon nuts or wedges are set, so that the preload remains in the tendon. By plotting the jack load against tendon movement, a load/elongation curve for the tieback is easily obtained. Our experience has been that these curves should be plotted for each tieback and they will have characteristic shapes for each job. The total strain will be the sum of anchor strain and tendon strain in its unbonded length.

The second test we use helps evaluate the anchor and unbonded tendon strain. We call it a performance test and use it only on a few tiebacks at each site. Normally, the first three tiebacks are performance-tested. We use a jack and gauge as for the proof test. The difference is that we systematically apply the loads in a pattern, so that the test load is applied in progressively higher increments, then decreased to zero in the same steps, then increased to the next-higher increment through the same steps. When the load is decreased, we of course get elastic rebound. This can occur in both the tendon and anchor, but experience shows that most of it is tendon strain. From this, we can verify that our tendon is unbonded for the proper length. Not all of the tieback movement is elastic. When the anchor is loaded, a portion of its movement is nonreversible. The performance test gives a good indication of the magnitude of this movement. By cycling the loads in this way, we can separate elastic from residual movement, and our experience is that this test is the most significant in evaluating tieback performance. By using only the movement the first time a particular jack load is applied, we can plot a load/deformation curve for the tieback. This curve is the most reliable method of predicting the pattern of movements to be expected with a proof test. We therefore use the results of the first three performance tests as a control for the proof tests. If a proof-test result is significantly different, the next tiebacks should have new performance tests performed. We also use the performance tests to estimate anchor movements. For most types of anchors, the residual movement from the performance test will represent the movement of the anchor at the end of the unbonded length.

We believe in field testing every tieback installed. Chapter 4 will discuss these tests in more detail. At this point, it is important to stress that these tests have been very effective in establishing reasonable patterns of tieback movements to be expected, identifying before their use in the projects tiebacks which will not carry at least a 120 percent overload, and modifying the original anchor design in the field when it proved inadequate. The results are too effective not to follow this procedure.

The design of tiebacks requires an understanding of the techniques available. We then select a technique which will give the design load required. The tendons and anchor are sized for this design load. The location of the anchor is checked to be sure it is far enough from the structure that its capacity will not be reduced by structure movement. The mass of soil which will act as a free body with the structure is also checked for stability. The anchor is relocated, if necessary, and rechecked. The number and spacing of tiebacks is selected based on the capacity of an individual tieback. The factor of safety is reviewed with particular emphasis on the reliability of the design load, and the design load adjusted accordingly. Finally, the individual tieback is tested to an overload, then preloaded. Tieback design results have been excellent using these general procedures.

3

CONSTRUCTION

Tieback construction techniques have been rapidly improving, and, as they have become more sophisticated, they have also become more economical and dependable. In most countries of the world, foundation specialists have led the way in these developments. In many cases there is a distinctly national pattern to the developments. For example, most of the German companies combine grouting and extraction of a driven casing to make a very excellent tieback. Sophisticated multiple-grouting procedures have been most carefully developed by French firms. The large hollow-stem-auger rigs are an American development, as are the belled tiebacks. This chapter will discuss some of these techniques for the four basic types of anchors of the preceding chapter.

Tiebacks with Straight-Shaft Anchors Made without Grout Pressure

The first tiebacks had straight-shaft anchors made without grout pressure. The techniques used a conventional drill to make the hole; then a tendon was placed in the hole and grouted. Rock drills were often used to drill through soil, then into rock. Progress was slow. On one project in Washington, D.C., three tiebacks a week was the average progress with each drill. A pipe had to be placed through the earth portion, but

FIG. 3-1 Rock drill.

even so the rock socket often became full of water. A drill such as that shown in Figure 3-1 was used, but it soon became apparent that better techniques would have to be developed if the tiebacks were to be economical in mixed-face soil.

A drill of this type is still most often used to drill tiebacks where the entire hole is in rock or the overburden has good drilling characteristics. The bit is a rotary percussion-type bit, which drills by shattering the rock. The hammer, which can be seen in the picture of the drill, is widely used for drilling rock holes, chiefly for blasting. This drill is a stock item which can be rented in most American cities. Lengths of drill steel are added with couplers, so that the hole can easily be drilled 40 or 50 feet into the rock. The cuttings are normally removed from the hole by blowing air down through the center of the drill steel and out through holes in the bit. The air then returns to the surface through the annular space around the drill steel and carries the cuttings with it. The air and cuttings normally carry out any water which seeps into the hole during the drilling. When the hole is drilled to its planned depth, air is blown through the bit until the air returning to the surface is clean. Then the bit and drill steel are removed from the hole. A tendon is inserted, and the anchor length is grouted. We prefer to insert a grout pipe to the bottom of the hole with the tendon and pump neat cement grout through it. The grout should be a stiff mixture of cement and water. The admixtures generally used to alter grout properties are seldom effective in this application and should be used with extreme caution. For example, most admixtures designed to cause expansion release hydrogen gas, which may be damaging to the steel tendon.

Construction techniques can seriously affect the capacity of this type of anchor. The grout is placed in the drilled hole and is dependent for its anchor capacity on bond with the rock. If the sides of the hole are covered with soft material deposited by the drilling technique, of course the bond will be low. Frequently rock drills can be used with soap or water to remove the cuttings instead of air. However, this is more likely to smear the sides of the hole and reduce anchor capacity. To get good capacity, the sides of the hole must be clean. If the grout shrinks when it sets up, the bond capacity will also be low. The best way to overcome this is to pump grout with a low water-cement ratio. A third problem that we have encountered with this type of anchor is flowing water carrying away the cement through fissures in the rock. If the tieback is converted to a pressure-grouting technique, the cement penetrates the fissures until it plugs them. Another method is to drill the hole, then fill the entire hole with grout. The next day, the hole is redrilled and filled with water. A pressure test is then made to evaluate the watertightness of the hole. If the test is satisfactory, the tendon is inserted and the anchor length grouted. If not, the grouting and redrilling are repeated until it is. The keys to making this type of rock anchor are drilling a sufficient length in the proper rock and making the anchor in a clean hole with nonshrink grout that will not be washed away.

Most tiebacks anchored in rock must be drilled through some soil. This has led to the development of a number of special techniques. The rig in Figure 3-2 has been specially developed to drill through soil with a two-tube drilling technique. The outer drill pipe is advanced only to the rock surface. Water is pumped through the inner drill pipe and carries the cuttings to the surface through the annular space between the two drill pipes. The soil may be cut by a variety of techniques. Frequently we use a casing bit with carbide inserts for the outer drill tube. We use a drag bit or roller cone bit on the inner casing to drill soil. When the outer tube is sealed in the rock, only the inner drill string is used to advance the anchor hole in rock. Normally this rig uses a down-the-hole hammer or roller cone bit to drill the rock. The cuttings have to be removed from the hole by a fluid, and considerable experimentation is still continuing on the best technique. Water, air, or a mixture of cement and water are now commonly used. When the anchor hole is drilled to its full depth, a tendon and grout pipe are inserted. The tendon is placed in the center of the rock socket and low water/cement-ratio grout is pumped down the grout pipe to make the anchor. After grouting the anchor, the outer drill pipe is also removed.

FIG. 3-2 Overburden drill for rock anchors.

Some engineers want the tendon to be permanently protected by a pipe through the soil. This can be placed around the tendon before the outer casing is withdrawn. In this way, holes can be drilled through earth into rock more rapidly than with a conventional rock drill, and special design features are more easily handled.

Other specialists use drilling techniques they have developed to drill for anchors in rock. Most anchors in rock are straight-shaft anchors made without pressure grouting because the grout-rock bond is high. Drilling speed is the key to economy, and most of the improvements on this type of tieback have been attempts to increase drilling speed. Straight-shaft anchors made without grout pressure are also used in earth, but not often.

In earth, straight-shaft anchors without pressure grouting are common only in soils that will stand open without support. In the United States, caisson-type drills have been used to drill holes. Figure 3-3 shows the type of rig most commonly used in this country. A short auger is attached to the end of a long "Kelly bar." The Kelly bar rotates the auger and also moves it into and out of the hole. The auger is drilled

FIG. 3-3 Caisson rig drilling tiebacks.

into the soil until it is filled with loose soil, then removed from the hole with the soil. After the soil is cleaned off, the auger is reinserted in the hole and drilling resumed until it penetrates enough to fill with loose soil again. Then the auger is again removed from the hole. In this way the auger is used to progressively lengthen the hole and remove the soil. When the hole is drilled to the desired length, the auger is removed, the tendon placed, and the hole filled with concrete or grout for the anchor length. The hole is about 12 inches in diameter, and the tendon is centered with steel "spiders" to keep it in the middle. Concrete is pumped to the end of the hole, then flows back up the hole toward the surface. Many anchors have also been made by pouring concrete down the hole, but with less satisfactory results. The resulting anchor is often 1 foot in diameter and 30 feet long.

Other tiebacks are made with continuous-flight augers which are drilled to the length of the tieback, then removed from the hole. The tendon is then inserted in the hole and the anchor length filled with grout or concrete. Figure 3-4 is a picture of one type of rig used to make tiebacks in this way. Sections of auger are connected together and drilled into the ground. Then the auger is removed from the hole in the reverse procedure. Straight-shaft anchors without pressure grouting are usually made in clay soils which can be drilled, have the tendon inserted, and then the anchor poured.

FIG. 3-4 Auger drill.

Tiebacks with Belled Anchors Made without Pressure Grout

The second general classification of tiebacks is belled anchors made without pressure grouting. These are made by augering the hole to its proper length with a drill such as the one pictured in Figure 3-3, then removing the auger and using a caisson belling bucket. The belling bucket has cutting arms which extend out when a force is applied to the bucket by the Kelly bar. The drill must be capable of simultaneously rotating the belling bucket and applying the necessary pressure. The belling bucket scrapes soil off the sides of the hole and collects the loose soil in the bucket. It is removed from the hole each time it is full, then reinserted, and the process repeated. Figure 3-5 shows a portion of the drill in Figure 3-3 with the belling bucket attached. This bucket excavates a bell-shaped enlargement, and, because of the way it is constructed, the bell diameter can be no more than three times the shaft diameter. The bucket will not remove all of the soil it excavates in the bell, so we normally use a hoe-type attachment on the end of the Kelly to drag out any loose soil which collects on the bottom of the bell. Some of this loose soil can also be pushed to the end of the hole without affecting the bell capacity. When the hole is belled and cleaned, a tendon is inserted and the anchor poured. The usual way of pouring this type of anchor is to pour concrete directly into the tieback hole and let the concrete run down the hole into the bell area. Naturally, this must be concrete which will flow on the angle from the horizontal at which the tieback has been drilled. We have also pumped concrete into this type of anchor, but the results do not seem to be any better.

Another type of belling bucket has been used in various places to excavate a different shaped bell. In general, this bucket has hinged

FIG. 3-5 Belling bucket.

cutting arms, so that it cuts a bell which is twice as long and shaped more like a nun buoy than a bell. These buckets have been developed by tieback specialists and are used only by their developers. Some of these buckets are activated hydraulically and can be used to excavate multiple bells.

All of the tiebacks discussed so far can only be used where an open hole which will stand open without any support can be drilled. The earliest tiebacks were made with these techniques, but today these techniques are used for only a small fraction of the tiebacks installed every year. Pressure-injected grout techniques have revolutionized the business and led to many new techniques.

Tiebacks with Shaft Anchors Made with Pressure Grout

Pressure-injected tiebacks probably started with Karl Bauer in Germany. He made tiebacks in sand using grout injected under high pres-

FIG. 3-6 Driving casing for pressure-injected tiebacks.

sures. A patent was granted, and his firm developed specialized tools and equipment to drive the casing and grout the tieback. Skilled crews were trained and developed experience with the technique. A network of companies in various parts of the world acquired the ability to install this type of tieback. New equipment, testing techniques, engineering understanding, and specialized accessories have led to steady increases in the use of this basic technique. Today, this is the most frequently used technique for making tiebacks.

This type of tieback is usually used in granular soils. The first step in making it is to drive a casing into the soil (see Fig. 3-6). The casing is threaded on both ends. Most common diameters are 3 and $3\frac{1}{2}$ inches. A starter piece is threaded onto the first piece of casing to be driven. This is a piece which has been machined to receive the point which is used to close the end of the casing. The point is attached, and the first length driven. The hammer is withdrawn, and another length of casing screwed onto the first length driven. This is then driven, and four or five lengths added in a similar way to result in a casing about 40 feet long driven into the soil. A tendon is placed in the casing. Hydraulic jacks are used to extract the casing as grout is pumped. The tight fit between casing and soil allows high pressures on the grout. Before grouting starts, the casing is pulled a short distance, and the point knocked off with the tendon. This opens the bottom of the casing to a

FIG. 3-7 High-pressure grout pump.

cavity in the soil. Then grout is pumped down through the casing using pressures from 100 to 600 lb/in^2. The grout hose is connected to the end of the casing, and, as the casing is jacked out of the soil, pumping continues. The result is a small-diameter anchor with very high capacity. The length of the anchor is determined by the high-pressure grouting. The last pieces of casing are pulled without pumping grout. The grout is a cement-and-water grout. When the anchor is made in porous soils, the cement-water ratio is unimportant, and ease of pumping is more important. The grout is mixed in a paddle-type mixer, then pumped with a positive-displacement piston-type grout pump (see Fig. 3-7). Grout behavior depends on the soil type in the anchor length. In porous granular soil, the grout can be pumped at high pressures, but the quantity required will be small. If the soil is clayey, when the grout hose is disconnected or the casing pulled, the grout is squeezed back out of the hole. Pressure-injected anchors do not perform well in clay soils for reasons discussed in Chapter 2. By observing the grout intake and backflow, we can often know the anchor is being made in clay. Then we change the tieback length or dip angle if this permits the anchor to be made in sand or gravel.

Pressure-injected tiebacks made with Dr. Bauer's basic discoveries are the most common type in use today. They are most effective in sand, gravel, and other porous soils where the high grout pressures create permanent high radial stresses in the sand. A number of other construction techniques are used to make shaft-type anchors by injecting grout under pressure. Any attempt to discuss all of these construction techniques would be too lengthy, but two typical examples will be discussed because they are based on drilling techniques and illustrate the basic approaches.

Some pressure-injected anchors are made in a drilled hole. A tendon is inserted into the hole, and a porous bag, which surrounds the tendon, is pumped full of grout to seal the hole. When the bag is first filled with grout, it presses against the soil surrounding the hole. As pressure is maintained, water from the grout escapes through the bag until the remaining cement becomes a solid. Then the anchor portion of the hole is isolated, so that as grout is pumped into the anchor length a pressure buildup can be accomplished. Small quantities of grout are then pumped into the anchor length to maintain this pressure. At the same time, water is passing from the grout into the soil and leaving the cement particles in the anchor zone. This process is continued until no more grout can be pumped at the desired pressure. Special grout pumps have been developed which permit both the pressure and rate of flow to be adjusted. This type of equipment is all made by and for the companies which have developed the particular tieback technique. The completed anchor made with this technique is similar to that made with the Bauer technique. It is usually about 4 inches in diameter and 20 feet long and designed for 50- to 100-ton capacity.

Hollow-stem augers are also used to make straight-shaft anchors with pumped grout. The auger diameter is usually larger than the two preceding techniques, and the grout pressures lower, but the resulting anchor capacity is comparable. The soil-grout contact area is larger, but the unit adhesion is lower. The rig shown in Figure 3-8 is used for this type of tieback. The key element is the long, continuous auger. The auger is screwed into the ground, and, as it turns, the hole it forms is the diameter of the auger. The soil is loosened and conveyed to the surface by the auger. If grout is pumped through the auger as it is withdrawn, then by coordinating pumping and withdrawing the auger, the anchor can be made in a cylindrical cavity. The auger and the soil on it confine the grout in the cavity and allow pressure to be used. They also keep the hole from collapsing during the drilling process. The auger should be as long as the tieback, so that the tendon can be inserted in it before the hole is drilled. We also avoid joints in the auger because they generally allow water to escape from the grout. Water escaping causes the grout to stiffen and harden, grouts the tendon to the auger, and prevents the flow of grout into the cavity. The grout is also different. In the other pressure-injected techniques the grout is usually cement and water, but we use sand and pozzolon in the grout

FIG. 3-8 Hollow-stem-auger tieback drill.

for this type of anchor. The grout can be truck-mixed. The grout is also pumped in a concrete pump which is not capable of the high pressures used with the preceding techniques. The technique we favor is to put the tendon, either a bar or strands, into the hollow stem of the auger. Then the auger is drilled into the soil to the desired depth. We are careful to leave enough soil on the auger to support the hole. Truck-mixed grout is pumped through the concrete pump and a hose to the end of the auger motor. It passes through a hole in the motor into the hollow-stem auger. It flows down through the auger and into the small cavity at the end of the auger. As pumping continues, a pressure build-up occurs. The auger is then withdrawn, and the tendon is left in the center of the grout anchor. When the auger has been withdrawn the length desired for the anchor, grout pumping is discontinued. By reversing the auger so that it carries soil into the hole, the rest of the hole can be backfilled with soil. The anchor made with the hollow-stem auger is a shaft in which the grout is pumped under pressure. We find this technique particularly economical in soils of stratified deposits.

More than half of all tiebacks are made with one of the techniques for making straight-shaft anchors where grout is pumped so that a pressure is applied to the soil surrounding the grouted anchor. The principles which make this procedure so effective in elastic soils were discussed in Chapter 2, but the application of these principles by a large number of tieback specialists has led to many effective techniques. Each technique may require special drilling equipment, grouting equipment, or tubes to be inserted in the hole with the tendon. Usually these variations have been developed by particular organizations and refined on the basis of experience until they are dependable. No attempt will be made in this chapter to discuss the details of all these systems.

Tiebacks with Enlarged Anchors Made with Pressure Grout

Grout can be pumped into some soils to enlarge the anchor or at least cause enlargement along its length. This is the fourth general type of anchor we will discuss. In general, these anchors are made by inserting into the grout a tube, which can be later used to pump successive grout injections. Usually the tube has holes which are closed by valves so that, after grouting, the grout cannot escape back into the tube. As in the preceding class of tiebacks, several methods of construction will be described. The reader should be aware that these systems have been developed by particular organizations. The same principles have been used by other organizations to develop systems in which the details are different, but all use tieback anchors which are enlarged by pumping high-pressure grout against the soil.

Bachy, a French firm, has developed a tieback with an enlarged anchor which is made around the grout tube, after which the tendon is placed in the tube and grouted in place. They call this the TMD anchor. Figure 3-9 shows the tube. It is a 2½-inch specially machined steel pipe. The pipe has holes in it through which the grout will be pumped. It also has rubber valves over the holes, which serve as one-way valves, to permit grout to flow out of the pipe, but not back into it. The pipe deformations serve to bond the pipe to the grout and also keep the rubber valves from slipping on the pipe. This special tube is used in the anchor length. It is attached to a smooth pipe of the same diameter, which extends from the anchor to the surface.

The first step in making this tieback is to drill a hole about 4 inches in diameter and fill the hole with grout. The grout tube, consisting of the special pipe and its smooth extension, is slid into place in the grout. The grout is allowed to harden overnight. Then the first regrouting is started by inserting a regrout pipe with a special packer into the tube. Small grout quantities are successively pumped through each hole. The packer presses against the tube so that each hole in the tube

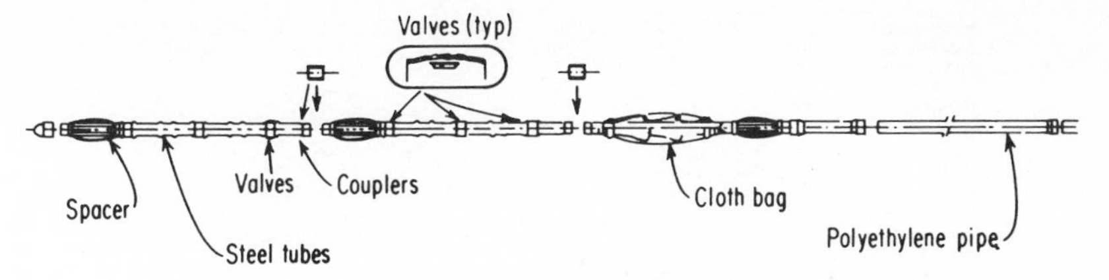

FIG. 3-9 TMD grout tube.

is isolated in such a way that when grout is pumped it opens the valve, breaks the grout around the valve, and displaces some of the soil around the anchor. The key to this procedure is being able to control which hole the grout is pumped through and then pump a controlled volume. After the first regrouting, the process can be repeated on successive days. Usually, the grout pressure required to displace the soil increases with each successive grouting. The capacity of the anchor is also increased as the regrouting is repeated. When the successive regroutings are completed, the anchor has multiple enlargements near each grout hole. The grout tube is flushed after each grouting. When the anchor is completed, the tendon is inserted into the grout tube and the grout tube filled with grout to complete the tieback.

This process uses the first grouting to seal the hole so the regrouting can apply pressure to the soil and not escape to the surface. The grout must then break the grout around the tube or it cannot escape from the tube. This requires high pressures since the grout hole is small, and special grout pumps with capacities of up to 100 atmospheres are used. As the grout comes through the hole in the grout tube, it spreads under the rubber valve. This increases the area of pressure on the grout until the grout breaks, after which a lower pressure is usually able to displace the soil. The pump is then run at slow speed until about 200 pounds of grout are pumped through a hole. When the pressure is reduced, the rubber valve closes over the hole, trapping the grout. It then sets up while the soil pressure is maintained. In this way, the soil can be pressure-grouted and its capacity to anchor the tieback improved.

The second type of tieback using an enlarged anchor made with high-pressure-injected grout has a regrout tube placed next to the tendon in the first grouting. The procedure is similar to the Bachy technique, except that the regrout tube is usually a smooth plastic pipe about an inch in diameter. As each grouting takes place, it breaks the anchor so it can escape from the tube. The section of an anchor in Figure 3-10 shows how the multiple groutings have done this. The

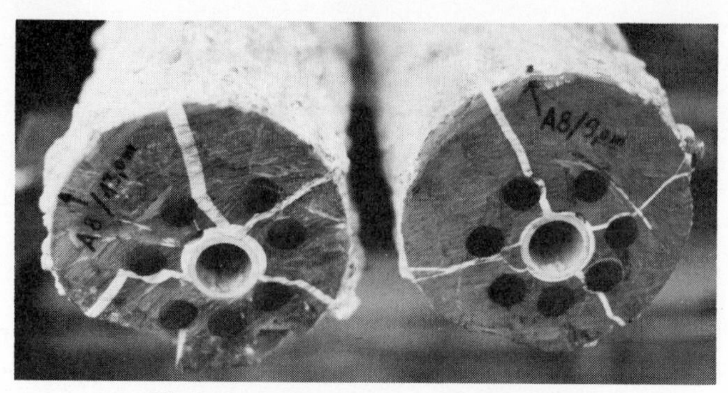

FIG. 3-10 Sections cut from regrouted anchor.

regrout pipe is clearly at the source of these anchor partings. The regrout tube has a series of holes, so that grout is pumped at discrete locations in the anchor length. As these multiple groutings take place, the size of the anchor is largest near each hole and smaller between the holes.

All of the anchors discussed so far use concrete or cement grout for the anchor. There are now two other methods we use for making anchors. In one, the grout is an epoxy. In another, the soil is hardened by the application of heat. Surely in years to come other methods of making anchors will be developed, but a brief description of these two is included to illustrate the potential.

We use the epoxy grout for straight-shaft anchors in rock. After the hole is drilled, sausages of the epoxy are inserted in it. These are about a foot long and an inch in diameter. The bar tendon is then turned and pressed into the hole. The drill hole must be about a half-inch larger in diameter than the tendon so that the bar rips the plastic container of the epoxy. The epoxy catalyst has a separate compartment in the sausage, and the turning tendon must also rip this compartment. Then, as the bar turns, it mixes the epoxy. The pressure forces the bar through all of the sausages until it reaches the end of the hole. We find a special chuck to mount the bar so it can be turned by the air-track drill is desirable since a great deal of torque (as well as thrust down the hole) is required to rip the sausages. The epoxy usually hardens in 5 to 15 minutes; so we continue turning the bar until the hardening epoxy begins to grip the bar. In a few minutes, the tieback is ready to test. We have only used this type of tieback in rock. The $1\frac{5}{8}$-inch-diameter hole can only be drilled economically about 10 feet; so the rock must have good bond characteristics. Where these desirable conditions are present, this is an excellent tieback, which can be used minutes after it is drilled.

We have only made a few anchors by hardening the soil, but the technique has a lot of promise. The strength of clays and some silts can be increased by increasing their temperature. Naturally, this also reduces their water content, which also increases their strength, if the temperatures are high enough. A heat source, in a small-diameter hole, can affect a sufficiently large mass of soil around the hole to increase the soil strength. In clay soils, if the heat input continues for a long enough time, a radial pattern of strength increases can be achieved, with the clay at the edge of the hole becoming bricklike. At this time this technique is experimental, but it holds great potential for future improvements.

Tendons

Tieback construction, in its simplest expression, is making the anchor and installing a tendon. The next part of this chapter will discuss tendons, particularly their placement. Later in the chapter we will discuss protecting the tendons from corrosion. In the last chapter we discussed the fact that all of the tendons currently used are some form of steel manufactured for use in prestressing, and as a practical matter either bars or multiple-strand tendons are used.

The tendon can be placed either before or after the anchor is grouted. In Chapter 2 we discussed the fact that the tendon is secured in the anchor, then extends between the anchor and structure. We provide a bond breaker on all of our tendons so that we can be sure they will be free to behave elastically. The bond breaker is placed on the tendon before it is installed. For bar-type tendons it is usually a smooth plastic tube, slightly larger than the bar, which is taped at each end and covers the bar in its unbonded length, as shown in Figure 3-11*a*. For strand-type tendons, a small grease-filled plastic sheath over each strand in the unbonded length is provided by the steel supplier. Figure 3-11*b* is a photograph of one strand tendon showing the strand covering in its unbonded length. Fabrication of the strand tendon also usually involves placing spacers and banding, so that the bond is increased in the anchor length, as shown in Figure 2-7. The spacers are stock items furnished by the manufacturer. We usually use a steel banding tool for gathering the strands together. Fabrication of the tendons may be more complicated in some of the proprietary systems, particularly when they are designed to be permanent, and some representative cases will be discussed later in this chapter.

Many tendons are simply slid into the hole before grouting starts. In the tiebacks made with anchors in unsupported soil, the hole is usually drilled, the tendon placed, and the anchor grouted. The tendon should

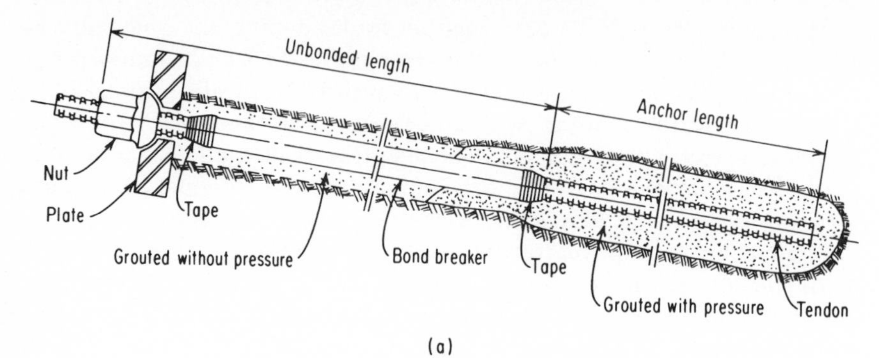

FIG. 3-11a Bond breaker on a bar tendon.

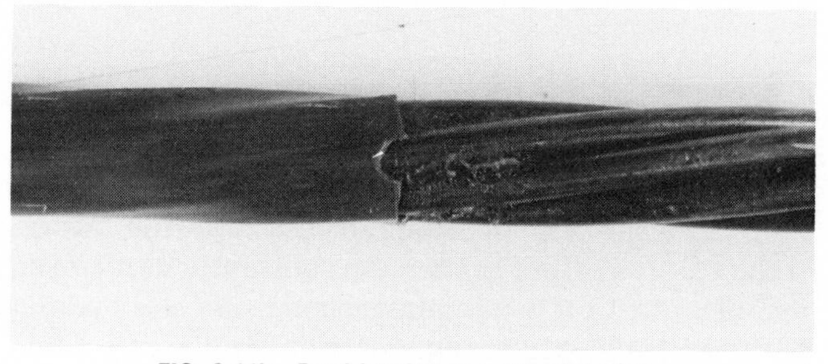

FIG. 3-11b Bond breaker on a strand tendon.

be held near the center of the hole with appropriate spacers. With the Bauer system, the casing is driven, then the bar placed in the casing. The point is driven off the casing by hammering on the bar, after which grout is pumped to make the anchor as the casing is pulled. With the hollow-stem-auger-type tieback, we usually place the tendon in the auger even before we start drilling. Most anchors in rock are made by inserting the tendon, then grouting the anchor length.

With the multiple-stage, pressure-injected anchor known as the TMD, the grout tube is grouted in place to make the anchor, then the tendon is installed. The tendon is installed by sliding it into the grout tube, which is first filled with grout.

Corrosion Protection

When tendons are corrosion-protected, special placement requirements must be met. Three common methods of corrosion protection will be discussed. These are the TMD strands, the Schnabel bar, and the compression-tube anchors. In each case, the tendon is encapsulated in an attempt to prevent air and water reaching it, so that it will not corrode.

In the TMD system, the grout tube is a special deformed pipe grouted into the soil for the full length of the anchor. A number of seven-wire strands are assembled to make the tendon. The tendon has no special corrosion protection in the anchor length, but depends for its protection on the fact that it will be surrounded by the pipe, which in turn has grout inside and outside. A predetermined quantity of grout is placed in the TMD pipe, and then the tendon shoved into the pipe. The amount of grout in the TMD pipe is the volume estimated to fill the anchor length. After the tieback is tested, the rest of the pipe is filled. The strand used is the same as that used for temporary tiebacks and described earlier.

Our company uses a different system for many permanent tiebacks. We use a deformed bar, which is coated with an electrostatically applied baked-on epoxy. The epoxy has been specially formulated to protect steel from corrosion. We then apply a special thermo-setting plastic pipe with an internal adhesive coating to the unbonded length. This is a plastic pipe which shrinks when it is heated. Heating also softens the adhesive so that the pipe shrinks and squeezes the elastic adhesive onto the bar. Over this, we then slip a smooth plastic pipe, which serves as our bond breaker. The first epoxy coating is a factory-applied coating for the full length of the bar. The heat shrink and bond breakers may be applied in the field. When all the protection is on the bar, it is placed in the casing or hole and grouted in place. This corrosion protection system is the only one compatible with the pressure-injected tieback developed by Dr. Bauer, and we have used it particularly with that system.

In Germany, a corrosion protection system known as the compression tube has often been used. The key feature of this system is a tube at the deep end of the anchor which causes the anchor to be in compression for its full length. Normally the tendon is attached to the anchor by bond, and, since the tendon is stretched, the anchor is in tension. This tension causes small cracks, or discing, in the grout anchor. The compression tube combines bearing and bond to apply the tendon force at the deep end of the anchor. The tendon is unbonded for its full length until it is attached to the compression tube. The corrosion-protected anchor is usually factory-assembled. The tendon is covered with a bond breaker. The compression tube, which is about 6 feet long, is attached to a smooth pipe of the same radius, and the tendon placed in it and screwed into the end of the compression tube. The space between the tendon and pipe is then filled with grout. This type of protection requires the tendons to be shipped to the jobsite after fabrication. A large enough hole is also required so that the assembly and any required regrout pipe can be placed in the hole.

Each of the corrosion-protected tendons I have described is a proprietary product. None of them is capable of use with all tieback construction procedures; in fact, they are really developed for use with particular systems.

This chapter has been devoted to constructing tiebacks. Representative methods have been described in some detail. The chapter is not supposed to be a comprehensive survey of all methods ever used. A careful reading should permit an engineer to understand the construction techniques which are successful. In Chapter 2, we discussed design; in this chapter, construction. The next chapter will discuss testing, which is easy and effective for tiebacks.

4 TESTING

Every tieback should be individually tested. In this chapter a specific testing program is recommended. A standardized testing procedure simplifies communication and interpretation of the results. This procedure combines proof and performance tests on every job and includes tests for creep. The tests are described in detail, and some discussion of test results is included. Following this standardized testing program, some special-purpose tests are discussed. A good testing program is essential since it is used to verify the design and modify it if necessary.

Each tieback is tested and usually preloaded. Preloading the tiebacks applies a force to the structure at the same time the tieback is strained. The preload strain in the tendon often exceeds the maximum tolerable movement of the structure. If the tendon only stretched and increased tieback load in response to structure movements, the tieback would not support the structure with acceptable movements. The preloading procedure is similar to prestressing. A hollow ram jack stresses the tendon, and nuts or wedges secure the tendon while loaded. The jack must be compatible with the steel used for the tendon and have a suitable capacity for both load and tendon elongation.

The Proof Test

Proof testing involves successively larger loads, so that a load elongation curve for each tieback can be plotted. This is the simple test used

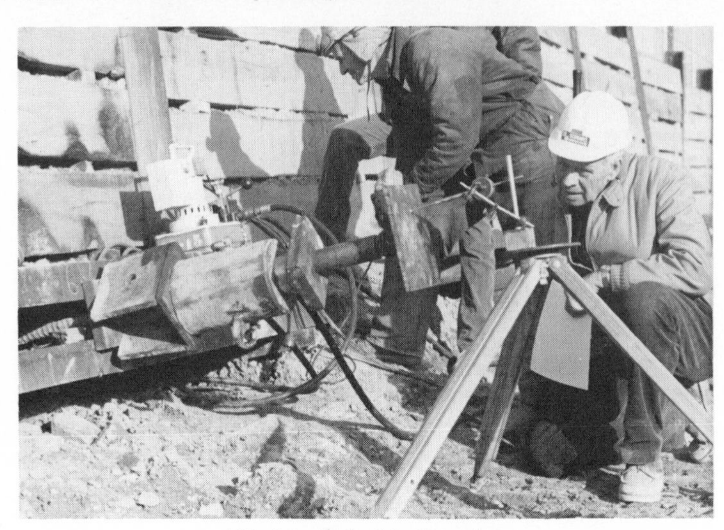

FIG. 4-1 Tieback testing setup.

on every tieback, except those that are performance-tested. After some experience, this test generates a great deal of confidence. Figure 4-1 shows a typical testing setup. Notice that the hollow ram jack slides over the tendon but rests on a "chair," which provides space to tighten the nut on the tendon at the desired preload. The jack is matched with a gauge and hydraulic pump and calibrated as a set. The load is read from the pressure gauge. A hollow circular-shaped load cell can also be placed under the jack to read the tieback load. We read elongation with a gauge designed to measure to thousandths of an inch. The gauge shown, mounted on a tripod not affected by movements of the structure, is most often selected. Another effective way of measuring movement is with a wire secured to the tendon. The wire is coaxial with the tendon, then passes over a pulley. A weight tensions it, and gauges measure its movement. With this procedure, the load/elongation measurements are obtained.

The proof test assures us that the tieback will carry its design load. In most cases, it is combined with applying a preload to the tieback. The tieback is first loaded to 2 tons, and the dial gauge read. Then the jack load is successively increased to 25, 50, 75, 100 and 120 percent of the design load, and the gauge read at each increment. If creep is not a concern, it is then reduced to the design load, then the final preload, which is usually 60 to 75 percent of the design load. The readings obtained are plotted on a simple load elongation chart, such as shown in Figure 4-2.

Elongation of the tieback results from straining the tendon and strain of the anchor-soil system. If the tendon is unbonded between the

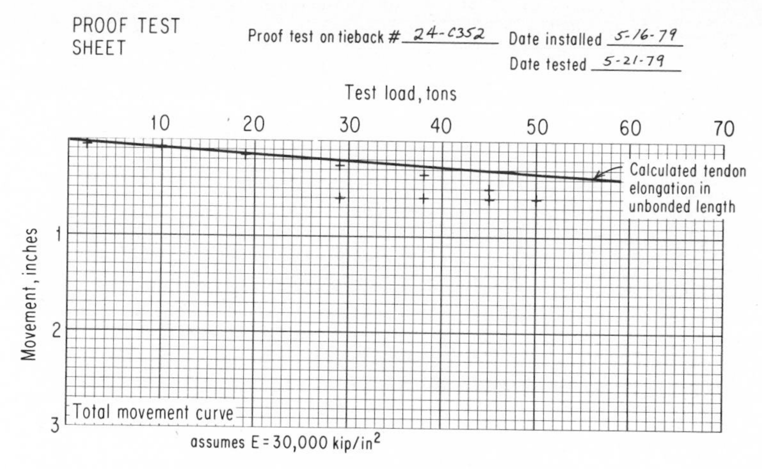

FIG. 4-2 Proof-test load/elongation chart.

structure and some part of the anchor, the strain of this portion of the tendon is easily calculated if we know nothing but the steel properties and the tieback load. The straight line plotted from the top left corner is the calculated tendon strain in its unbonded length. The vertical distance between this line and the load/elongation curve is the total elongation from all other causes, which include:

- tendon strain in the anchor length
- slippage at the tendon-grout connection
- strain in the grout anchor
- slippage at the grout-soil connection
- strain in the soil around the anchor

as well as possible errors in measuring load and elongation. Later in this chapter, when test interpretation is discussed, these components of tieback elongation will be discussed in more detail.

The Performance Test

The performance test is the most reliable way of predicting the load/elongation behavior of proof tests. It should be used on some tiebacks on every project, and all other tiebacks should then be proof-tested. The performance test should also be used for its value in understanding and separating the causes of tieback elongation.

Performance-testing a tieback involves load/elongation measure-

0 tons	.75 D.L.	.25 D.L.	.25 D.L.
2 tons	.50 D.L.	2 tons	2 tons
.25 D.L.	.25 D.L.	.25 D.L.	.25 D.L.
2 tons	2 tons	.50 D.L.	.50 D.L.
.25 D.L.	.25 D.L.	.75 D.L.	.75 D.L.
.50 D.L.	.50 D.L.	1.00 D.L.	1.00 D.L.
.25 D.L.	.75 D.L.	1.20 D.L.	1.20 D.L.
2 tons	1.00 D.L.	1.00 D.L.	1.33 D.L.
.25 D.L.	.75 D.L.	.75 D.L.	1.20 D.L.
.50 D.L.	.50 D.L.	.50 D.L.	1.00 D.L.

FIG. 4-3 Sequence of jack loads for a performance test. D.L. = design load.

ments in a particular sequence of increasing and decreasing loads. Jacks and gauges are similar to the proof-test setup shown in Figure 4-1. The sequence of jack loads is shown in Figure 4-3. Each load in the schedule is applied to the tieback in the order listed. Since it is difficult to pump the jacks to the exact load listed, the actual load should be recorded if it is slightly different. At each load, elongation of the tieback is measured and recorded. Figure 4-4 shows a typical record of jack loads and elongation for a performance test. The shading and cross-hatching have been added to help illustrate the use of the measurements. From this data, two load/elongation curves are plotted.

The first load/elongation curve plotted, the upper part of Figure 4-5, uses the shaded points. Note that, when these are plotted, the loads are the same as for the proof test, except that one higher increment of load has been added. The magnitude of elongation is the same as it would be for a proof test on that tieback. This can then be used as a visual prediction for proof tests on adjacent similar tiebacks. The only difference in the test procedures is that, for the performance test, the load is reduced to zero (2 tons is considered a nominal zero) after each shaded load is applied. It is then increased in steps to an increment higher than the previously shaded load. This cycling does not affect the magnitude of elongation recorded the first time each increased increment is applied. The shaded readings are for these particular readings. We have tested hundreds of adjacent tiebacks with proof and performance tests to confirm that the shaded readings are not affected by the cycling. Thus, one result of the performance tests is a load/elongation curve which predicts the load/elongation behavior of the proof tests.

We also plot the residual anchor movement from the measurements obtained. This is the lower curve of Figure 4-5. The elongation at each cross-hatched line is plotted against the highest load previously re-

Basis of load	Load wanted	Load read	Gauge reading
At no load	0T	0	000
At 2 tons	2T	2	041
Increment #1−0.25 x Design Load	10	10	088
2T	2	2	064
#1	10	10	092
#2−0.5 x Design Load	18.75	19	160
#1	10	10	109
2T	2	2	048
#1	10	10	192
#2	18.75	18	164
#3−0.75 x Design Load	29	29	271
#2	18.75	19	215
#1	10	10	140
2T	2	2	085
#1	10	10	125
#2	18.75	19	200
#3	29	29	289
#4−1.0 x Design Load	37.5	38	391
#3	29	29	350
#2	18.75	19	258
#1	10	10	162
2T	2	2	096
#1	10	10	132
#2	18.75	19	222
#3	29	29	311
#4	37.5	38	400
#5−1.2 x Design Load	45	45	469
#4	37.5	38	455
#3	29	29	383
#2	18.75	19	292
#1	10	10	180
2T	2	2	100
#1	10	10	145
#2	18.75	19	235
#3	29	29	325
#4	37.5	38	423
#5	45	45	480
#6−1.33 x design load	50	50	560 13
#5	45	45	538
#4	37.5	38	470
#3−Lock-off	29		420

FIG. 4-4 Typical record of a performance test.

corded. The plot is the movement of the anchor, which is not elastic, as a result of each load increment. Remember that elongation of the tieback results from strain in the unbonded length of the tendon plus the movement in the anchor. The movement of the anchor can also be thought of as divided into elastic and nonrecoverable movement. In Figure 4-6 the total movement of a tieback has been separated into these three portions. When a load is applied to the tendon, the anchor

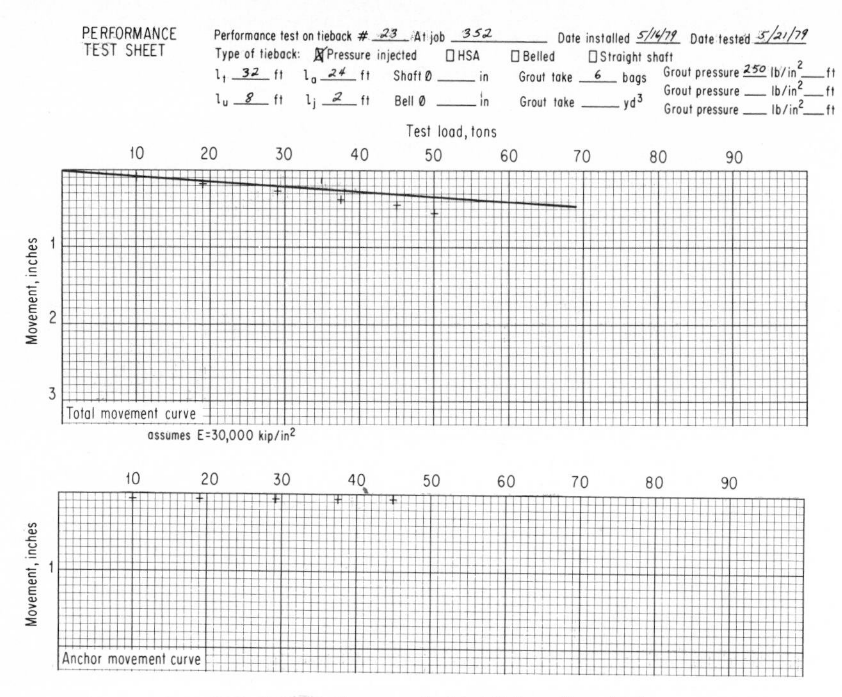

FIG. 4-5 Performance-test load/elongation chart.

strains. When the load is released, some of this strain is recovered, but some is not. At this stage of the state of the art, we cannot reliably calculate either component of anchor strain, but with the performance test we can directly measure the nonrecoverable anchor strain. We call this residual anchor movement.

This is a reliable method of measuring residual anchor movement. When there is no load on the tieback, there should be no strain in the unbonded length of the tendon. The measured elongation is the amount the anchor was permanently displaced by the loads which preceded the particular reading. We have done a large number of tests on tiebacks using extensometers, which verify that the anchor moves when loaded and partially rebounds when unloaded and that the residual anchor movement can be reliably measured with the performance test.

Residual anchor movement depends on the tieback construction and the soil. Straight-shaft anchors made in sand with pressure-injected grout, such as the Bauer method, move least. The work of Lyman Reese (1976) on machine-drilled piers has highlighted the fact that required strains to reach ultimate bearing capacity are much larger than the

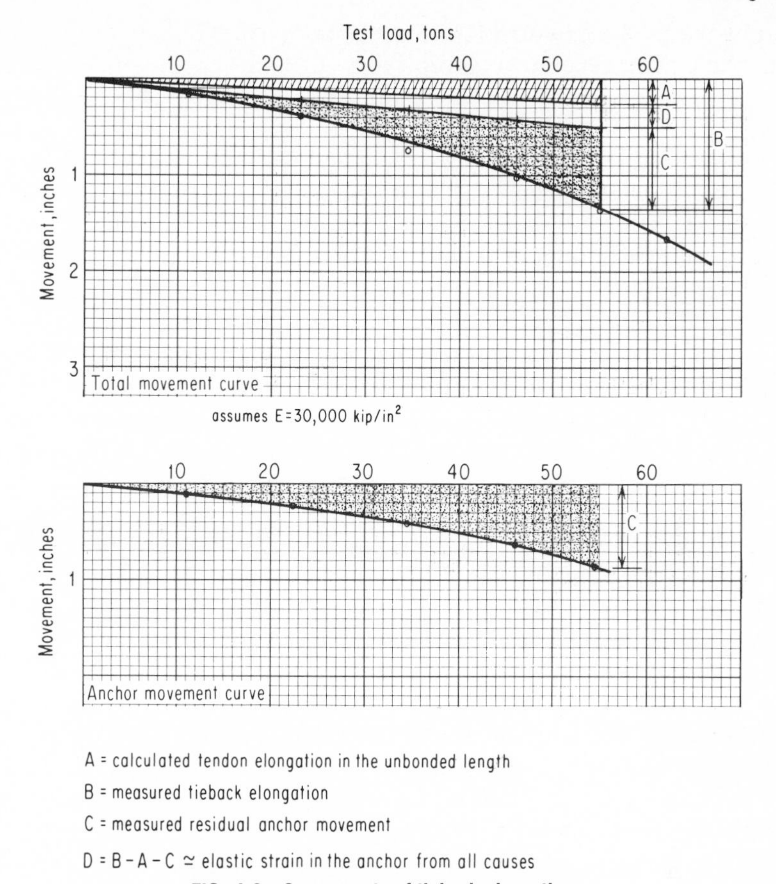

FIG. 4-6 Components of tieback elongation.

required strains to reach ultimate shear capacity in most soils. We would then expect belled anchors, which depend on bearing, to move more than shaft anchors which depend on shear. Many tests have confirmed this. Two examples will be discussed.

The tieback in Figure 4-5 was made with the Bauer technique in sand. When it was performance-tested, residual anchor movement at 45 tons was only about 0.10 inches. We have run a great many tests on similar tiebacks and expect about a quarter inch.

The tieback of Figure 4-6 was made with a belled anchor and no pressure on the grout. Residual anchor movement was about 0.9 inches. We have run a great many tests on similar tiebacks and have often used tiebacks where the residual anchor movement exceeded 2 inches.

More movement is required to develop anchor capacity by bearing than by shear. We can directly measure the residual anchor movement from the performance test. From tests with extensometers and strain gauges we also know the anchor rebounds elastically. These tests indicate that anchor elastic strain is approximately equal to the difference between measured tieback elongation and the sum of tendon strain in the unbonded length and residual anchor movement. Total tieback elongation with a belled anchor is often dependent on residual anchor movement, while with a Bauer anchor the tendon unbonded length usually controls.

The performance test is often used to indicate that the tendon is unbonded in the desired length. When the tieback load is increased, the measured elongation should exceed the strain in the unbonded length. When the load is decreased, the tendon rebounds. If anchor movement is small, which is the case in many pressure-injected straight-shaft anchors, tieback elongation and tendon strain in the unbonded length should be related. When we reduce the load to zero from each of the shaded loads of Figure 4-4, the tieback rebounds. If anchor elastic strain is small, tieback rebound should be equal to the tendon rebound. Then when the load is increased, the tieback should elongate because of tendon, elastic anchor, and increased residual anchor strain. If we next reduce the load and measure residual anchor strain, we can subtract this from tieback elongation to estimate the strain in the unbonded tendon length. The method can be used with a great deal of confidence when anchor movements are small.

If the tendon is unbonded in the desired length, we are sure it is not transferring load except in its anchor length. In Chapter 2 the need to place the anchor beyond the critical failure surface was explained. The unbonded tendon length is designed, among other things, to be in uniform tension between the structure and anchor. If some tieback load were accidentally anchored in this zone, the test would indicate capacity which could not be relied upon. The performance test helps verify that the anchor is far enough from the structure by confirming the tendon is unbonded in the planned length.

With the performance test results we can verify that the tieback will carry the desired load and make a reasonable determination of residual anchor movement and the unbonded tendon strain. We are then in a position to anticipate the movements associated with proof testing similar tiebacks. We can also predict the elongations of similar but shorter or longer tiebacks with anchors in the same soil. This latter fact is based on the fact that if anchor strain is small or a constant, tieback elongation is simply related to the unbonded length. This fact has been used by some engineers to establish acceptability criteria for tiebacks. These criteria are most effective for tiebacks with long unbonded tendon lengths and small anchor strains. They should not be applied indiscriminately to all types of tiebacks.

Anchor Strain

Some efforts have also been made to calculate anchor movements. Our experience is that these have only limited application. For example, the soil stiffness should be the controlling factor for a belled anchor. However, correlation between the calculated and measured movements is poor. Further research and study are required before an analytical calculation procedure can be recommended.

Movements of straight-shaft anchors are equally difficult to calculate with assurance. Fortunately, they are usually smaller, so they are less important. Numerous procedures have been recommended, but the interrelationship of the tendon strain in the anchor length, strain between the tendon and anchor to create the desired bonding, strain and cracking of the grout anchor, strain between the grout anchor and the soil to create the desired bonding, and strain in the soil all contribute to the movement.

Simplifying assumptions can be made to permit calculation of shaft-anchor movements. For example, we can assume that all stress is transferred to the soil uniformly and that the deep end of the anchor does not move at all. If the tendon is bonded to the anchor for the full anchor length, then the stress in the tendon decreases from the jack load to zero uniformly. Total tendon strain in the anchor can be calculated. It is equal to the strain calculated for the jack load and half the anchor length. Thus, using these assumptions, total tieback movement should not exceed strain in the unbonded tendon plus half the anchor length for any load. For a shaft anchor made in a uniform soil, this may be a safe approximation.

Measurements have shown that this is not a true model of anchor behavior. The tieback load is not transferred to the soil uniformly. In a uniform soil, the rate of load transfer is a function of strain. Thus, the transfer rate is a maximum near the front of a shaft anchor. Since this results in less tendon strain than calculated in the preceding paragraph, it is an error on the side of safety. If the shaft anchor is made in nonuniform soils, then the rate of transfer will be affected by the soils. If the soils at the front of the anchor are not as stiff as the others, then the rate of transfer will be lower at the front, and anchor movement can exceed that calculated in the preceding paragraph. If the transfer rate is a function of strain, but peaks for very low strain and then decreases with larger strains, the total anchor strain can exceed that calculated in the preceding paragraph. Some movement of the deep end of the anchor can also occur and would have to for a uniform rate of transfer. This must result in larger anchor movements than calculated in the preceding paragraph.

The present state of the art does not allow us to calculate anchor movement confidently. The performance test, by measuring residual anchor movement, allows a better estimate of anchor movement.

Creep

The performance test assures that the particular tieback method will work at a particular site and gives an acceptable load elongation pattern. In all of the preceding discussions, creep has been ignored. For tiebacks in rock and granular soils, creep is no problem. In soils where creep is a consideration, it can also be evaluated as a part of this test. Some changes in the test setup must also be considered so that the test load can be maintained carefully while tieback elongation takes place. Special jacking systems, using pumps which are activated by the small pressure changes resulting from creep, are used. Figure 4-7 shows such

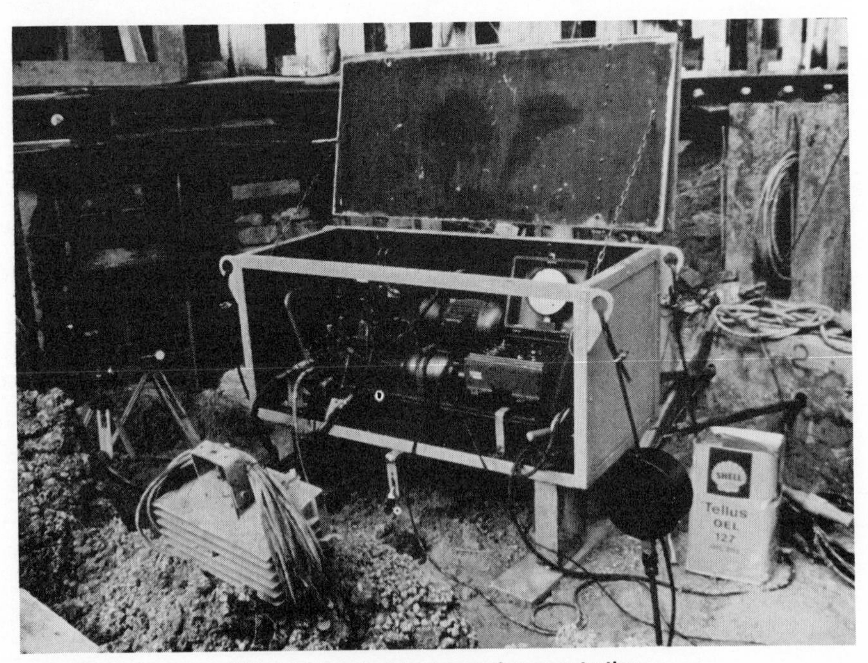

FIG. 4-7 Constant-load pump for creep testing.

a pump. One problem with this arrangement is that all test jacks have some friction, and, for the careful measurements needed to analyze creep, maintaining uniform pressure in the hydraulic system does not guarantee uniform load on the tieback. Load cells are available which can be used to measure the tieback load exclusive of jack friction and thus maintain the tieback load more accurately. The readout is very sensitive to changes in tieback load. The best combination is a load cell used with the special pump. Controls which allow slight adjustment of the pump pressure permit the tieback load to be monitored precisely and the hydraulic pressure to be adjusted to overcome jack friction. To check creep as a part of the performance test, the tieback load is maintained constant while the creep is carefully measured.

The measured movements are carefully analyzed to determine whether the creep is within acceptable limits. To properly analyze the creep, a pattern of time-dependent movement is studied. The procedure we use is to measure creep only for the design load. The schedule of time/elongation readings is shown in Figure 4-8 for the performance test. Note that the time starts when the pumping starts, and the first elongation reading is taken 1 minute later, by which time the load

FIG. 4-8 Schedule of loadings for creep testing.

Time interval	Time to read	Gauge reading	Creep
Before loading			
1 minute			
2 minutes			
3 minutes			
4 minutes			
5 minutes			
7 minutes			
10 minutes			
15 minutes			
20 minutes			
25 minutes			
30 minutes			
45 minutes			
60 minutes (1 hour)			

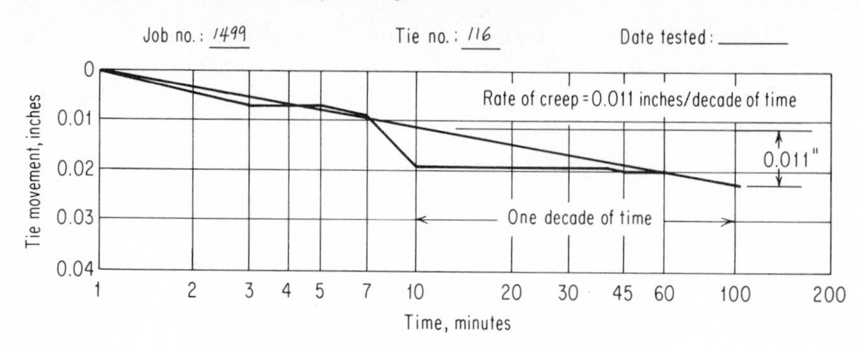

FIG. 4-9 Creep measured during performance test.

must be on the tieback. As the tieback creeps, the jack must be repumped to maintain constant load. At the selected times, the elongation of the tieback is reread and recorded. Since the magnitudes of acceptable creep are very small, the readings will be useless unless the load is carefully maintained. When all of the readings are completed, they are used to prepare a plot of time versus creep, as shown in Figure 4-9.

Creep of an acceptable tieback must decrease with time, and the total magnitude must not exceed acceptable limits. Figure 4-9 helps us verify both criteria. The test results are plotted on semilogarithmic paper, with time plotted on the logarithmic scale. Tieback elongation is plotted against the log of time. This also permits a reasonable estimate of total tieback creep. Measurements of creep have occurred over a period of 60 minutes, resulting in nearly two "decades" of time. The slope of the resulting curve is expressed in movement per decade. If the creep continues to decrease with time, but follows the same rate per decade, then we can predict it. Figure 4-10 shows the test results reported in Figure 4-9, but with various future times added to the diagram. Again, if the rate continues, we can estimate total creep of the anchor at any desired time. If the creep of the anchor agrees with this model, we can predict the creep which will occur at any time. The performance test can thus also be used to analyze creep.

The Test Program

We use the performance test on the first three tiebacks tested on any project. We also use it selectively on later tiebacks. If the job has a large number of tiebacks, another 1 or 2 percent should be performance-tested routinely. If soil properties change, we test additional tiebacks. If significant variations in the tieback construction occur, each variety

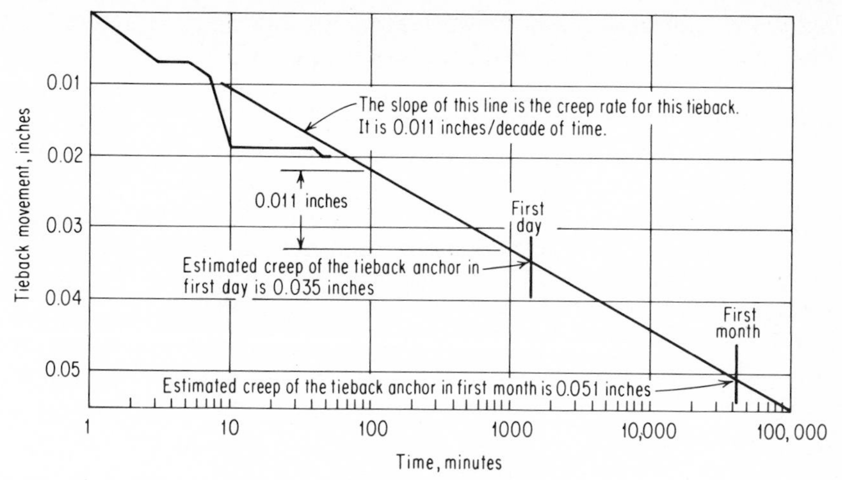

FIG. 4-10 Future times added to Figure 4-9.

of tieback should be performance-tested. The tiebacks which do not require a performance test are proof-tested. In this way, every tieback is tested before it is incorporated in the work. We know it will carry the design load and not creep excessively. By preloading it, we also preload the structure before any movement occurs. When all tiebacks are successfully tested, we have completed a project of which we can be very confident.

Not all tiebacks pass these tests. When a tieback cannot pass the performance or creep test, modifications to the construction procedure or a reduction in the design load per tieback usually result. When a tieback cannot pass the proof test, it should be analyzed, assigned a revised design load, and additional load capacity provided by other tiebacks. Only tiebacks which are dependable should be incorporated in the project.

The performance test assures us of the suitability of the particular tieback for the site. When a tieback fails the performance test, field modifications generally follow. Perhaps the anchor length is increased, the overall tieback length or inclination angle is altered so that the anchor is made in a more favorable soil, or the details of construction are changed in some other way to improve the performance. Then, the revised tiebacks are performance-tested. When the field modifications lead to acceptable performance, the same changes are made for the subsequent tiebacks. In this way, the performance test can be used to help analyze and improve the tiebacks. We make our original design based on a best estimate of available capacity. This is checked in the field with the performance test, and, if the capacity of each tieback must be reduced, additional tiebacks have to be provided. The performance-testing program is not completed until a procedure for constructing the tiebacks can be followed which should result in tiebacks capable of carrying a particular design load. The performance test is the final step in the selection of the tieback design.

When a tieback does not pass the creep test, field modifications also follow. Altering the tieback construction method, the location of the anchor, or the design load are considered. The changes are made, new tiebacks constructed, new tests performed, and this process repeated until successful tests have verified that the modifications necessary for successful performance have been made. The creep test is a part of the performance test in soils subject to creep. Changes are made until successful results are obtained. Then the changes are incorporated into all similar tiebacks. The performance and creep tests thus are used to polish the design at the start of the job.

The proof test is used to screen the project for tiebacks made with the proper technique which nevertheless will not carry the design load. If failures are numerous, additional performance tests and modifications are in order. The proof test, properly used, detects the random tieback whose failure can only be explained by localized soil conditions or other unexplained causes. It is used to prove that each tieback accepted will carry its design load. When one fails the test, it should be analyzed. Usually, it is derated based on a factor of safety, and an additional tieback furnished to provide a portion of the original design load. If all tiebacks are proof-tested, we can be sure that each tieback will carry the load assigned to it in the final design.

In this testing program, each tieback is tested. The result is that the tiebacks used can be relied upon. All of the testing discussed so far in this chapter should be done on every tieback project. In addition to these tests, there are several others which will be discussed in more detail in later chapters. An experienced tieback specialist should be familiar with them, but only a brief description will be included here.

Testing-Program Modifications

Tests to failure can be performed, but in tieback practice these are rarely used. Both the tendon and the structure are usually designed so that the maximum practical overload is 50 percent of the design load. This should not cause failure of most tiebacks. It is also difficult to extrapolate two or three tests to an entire project, so the combination of proof and performance testing is more reliable.

The overloads to which tiebacks are proof-tested can be increased. Twenty years of experience using 120 percent as the maximum test load for proof testing have been successful. Other overloads can be arbitrarily selected, but there is no engineering basis for them. The overload used is more than enough to compensate for tendon creep, jack errors, and other testing errors and assure that each tieback has been tested to more than its design load.

The time of creep testing can be extended. For example, 16, 24, or 48 hours are often required. This is an area in which our company is continuing studies. The period recommended is 60 minutes. One more decade (600 minutes) requires holding the load for ten times as long, or 10 hours. This effort cannot be justified by any experimental data at this time.

The number of load increments at which creep measurements are made can be increased. This procedure is used by some specialists to analyze creep. As the load increments are increased, creep is measured. Then, these measurements are analyzed to determine a critical load below which the creep is satisfactory. These tests must be performed well before production tieback installation to be effective. If a suitable tieback can be determined without this test and successfully tested using the recommended test, this is not necessary. Creep can be successfully analyzed in most cases without this test.

Other special-purpose tests are occasionally made. Load cells can be placed on tiebacks to measure tieback loads over extended periods. Tiebacks can be retested at various times after completion.

Conclusion

The tests recommended in this chapter—performance and proof tests—are general tests that should always be used. In most cases, if these tests are carefully done, no other tests are required. They will assure the ability of each tieback to develop its design capacity. The mass which is then reinforced and tied together must also be checked for stability, as in Figure 2-6.

Tieback testing is both a control on quality and a help in design. The program of performance testing, evaluation, design modification, and proof testing recommended has been completely successful. If this program is understood and followed, it will facilitate the comparison of results on various projects. It also simplifies the communication of results. Slight variations which alter test procedures may also produce differences in the results. The variations and their effects must be understood to compare the results. If standard procedures are used which eliminate these variations, a body of useful experience can be more rapidly developed with the tests. Good control tests should also be consistently performed, and for that reason they have been described in detail. This recommended testing program should be performed on every tieback project.

5 CONTRACTING PROCEDURES

Contracting procedures have as much influence on the use of tiebacks at this time as any other single factor. For the reader who is considering using tiebacks, it is as important to understand the contracting procedures as the design, construction, and testing procedures. At this time, the most successful procedure is to acknowledge you are buying a specialized product from a specialist and concentrate on the performance, not the details of construction.

A number of contracting procedures have developed, which yield successful applications. Most tiebacks in the United States have been used on private temporary work, and the contracting procedures for this work will be reviewed. In Germany, a different procedure is used, and this will be discussed. The French use still another procedure but also obtain good results. Finally, some discussion of problems with the usual American public-works procedure will be included. The goal of this chapter is to allow the reader to use tiebacks with confidence.

Any good contracting procedure must result in the selection of a tieback design that is appropriate for the soil and the intended use, has no adverse effects on utilities or structures, is economical, and can be installed in a reasonable time. The key word is "selection." Select only systems which meet these criteria. Contracting procedures which allow tieback selection from among those with a proven record of success should be favored.

The Normal American Contracting Procedure

Here in America most tiebacks have been used to support excavation bracing walls. The normal contracting procedure for this has been to make the contractor completely responsible for both design and construction of these walls. They are temporary structures, needed only during the construction phase of the project, and the scope of the work required is often affected by other decisions the contractor makes. With this procedure, proprietary products can easily be used, and as a result almost all tiebacks used for this work are proprietary. Complete freedom of choice by the contractors has allowed them to develop effective, efficient systems and use them. Analytical tools are readily available which permit other engineers to review the contractors' design and test the capacity of each tieback, so that improved methods can promptly be utilized. This performance-demanding climate is the stimulus for most of the American tieback industry, so let's examine it a little more closely.

The first step is to establish that an excavation bracing wall is required. These walls are usually constructed to protect streets, alleys, and adjacent properties. When deep excavations are made, usually the owner of the project or contractor determines that the protective wall will be required. Deep excavations requiring excavation bracing walls are usually in the heart of the city; in industrial plants; or for subway, sewage, or highway projects. They are a construction expedient to permit the permanent structure to be built.

When a tied-back wall is needed, the contractor usually solicits proposals from specialists in this work. Typically, he expects these firms to be able to assist in establishing the scope of the wall design and construct it, monitor the work, and make minor modifications to adjust to field conditions as construction proceeds. In most cases, the contract is a simple performance requirement, usually a lump sum design and build contract.

Design involves identifying the earth pressures which will act on the wall, then developing the number, capacity, and spacing of the tiebacks required. The wall is designed to span between tiebacks. The tiebacks are selected so that they are appropriate for the soil and structure and designed to carry the required load. The soil in which the tiebacks are anchored should be deep enough that they will not pull out if wall movement occurs and also long enough to assure mass stability. All of these design considerations are usually analyzed by the specialist as a part of his contractual obligation. For this type of contracting procedure, an engineer in the specialist firm makes the complete design.

He is responsible for selecting a tieback design that will perform. He is also knowledgeable about the advantages and limitations of the methods his firm has used. In earlier chapters we discussed a representative selection of methods for making tiebacks. No one firm is able to make every type of tieback. Obviously, patents would cause this, but specialized techniques and equipment also lead to the same result. Each firm that specializes in tieback work has concentrated its efforts on developing particular tieback methods. As soon as we perfect a method, we can use it. The methods we use have resulted from decisions to acquire specialized tools, equipment, or techniques; careful training of crews; and investigations to understand the method. We must continuously exercise judgment to be sure we are using the most efficient methods possible.

This procedure recognizes the selection process. Accepting this as a desirable part of tieback design is one of the keys to the success of this particular contracting procedure. Only methods with a record of successful application survive. The user should let only experienced engineers choose methods which are appropriate for the soil and should insist on a valid testing procedure. Finally, and this is very important, this contracting procedure has gained acceptance because it is not only more economical, it has resulted in better performance.

This should not be too surprising. Tieback specialists have everything to gain from good projects. With control over both design and construction, they have accepted responsibility for good performance. The result has been the development of a number of tieback specialists who do almost all the tieback work in America. These firms have accepted the responsibility of developing a new type of construction and have a vested interest in its success.

In many cases, the selection of the tieback method is made at the jobsite during construction of the project. In stratified soil, a prudent designer will recognize the possibility of using several methods. He will allow the selection to be made between acceptable methods when the soil in the anchor length can be identified. This may involve several kinds of equipment on the same project, but it is foolish to try to make a tieback method work which is not suitable for the soil.

This American contracting procedure for tiebacks has been very successful. It has been used almost exclusively by private owners for excavation bracing walls and most public agencies in America. It has stimulated innovation, because it allows specialist companies to incorporate in their bidding and design the advantages of new methods. It has encouraged them to study the methods in use and to staff and find programs designed to improve on these methods. The result has been the development of the American tieback industry.

European Contracting Procedures

In Germany, a more formal recognition of the proprietary nature of most tiebacks is the central feature. All tiebacks, temporary or permanent, must be installed in accordance with a national standard. The essence of the standard is that each tieback method used by any firm must be prequalified. The prequalification is issued to a particular firm for installing a particular kind of tieback in specific kinds of soil. The contractor, to obtain the prequalification, installs test tiebacks with the particular method in the appropriate soil. The installation is inspected by a foundation institute. The tiebacks are then tested and exhumed. A report is prepared by the foundation institute which details the construction procedure followed and the test results, and recommends capacities. The final approval is certification to that firm for that capacity and type of tieback in that particular type of soil. Several organizations can, of course, seek and obtain approval of similar kinds of tiebacks, and each organization must individually seek approval of any tieback method it intends to use. The project designer then simply requires an approved tieback or selects from those previously approved. If several methods are desirable on the project, he is free to use the proper mix. This formal procedure usually allows the choice to be made on the basis of economy by permitting competition in Germany between all firms which have been determined in advance to be competent to install tiebacks in the particular conditions. The firm selected has complete specifications based on their prequalification and of course has skilled manpower, special equipment, and experience. They must also handle the details to make the tieback selected work. It is their system.

The French use a contracting procedure which also places heavy reliance on specialty contractors and jobsite tests. The choice of the method and design of the tieback using that method is up to the tieback contractor. The code deals particularly with the materials used to make the tieback, prestressing steel, cement, and corrosion-protection materials, and the testing procedures. Tests at the jobsite are required prior to approval of the design for particular conditions. The contractor may be required to make a number of tiebacks at each site using the method he has selected for the production tiebacks. Based on the results of these tests, he may have to modify the tieback design and install additional test tiebacks. No work on production tiebacks can begin until this program of testing has been satisfactorily completed. Then during the installation of the production tiebacks further testing is required. The key feature of this French contracting procedure is, again, the specialists who have broad discretion in the choice of tieback method and design, using the method selected, but requiring their tiebacks to pass a performance test.

Except for details, we can see that there is a great deal of similarity in the American, German, and French contracting procedures. In each case, specialists do most of the work, with methods they have developed. Each procedure has encouraged these specialists to develop economical or improved methods, train the men, secure the equipment, and procure the supplies and tools that will steadily improve their performance. Competent engineers, working in a specialized field, have developed tieback methods which dominate the industry in each country. In each country, the methods used have been developed over a long period of time, and are peculiar to the country. The key to the contracting procedure is to recognize that tiebacks are systems which are available to use. To use them successfully, the project designer must know that the tieback firm is capable. The Germans do this through prequalification. In the United States and France, judgment is required. Proposals should be solicited only from qualified firms, but the firms should have maximum freedom to design the work. After the prices are in, be sure the low bidder's design is compatible with all other needs of the project and award the contract. The tieback test procedure should be clear. Performance should be the responsibility of the specialists, and such design decisions as tieback capacity, length, etc., should be made by them. If replacement tiebacks are required, the specialists should pay for them.

This procedure can be modified when it is desirable. For example, additional criteria, which the contractor must design for, may be established. The earth pressure which the wall must be designed to support may be established. Often, a slope up from the top of the wall to existing grade is planned, and the allowable slope is specified. From these examples, it should be clear that there may be cases for which the contractor should be asked to satisfy additional criteria. In general, these should only be included after careful consideration. More tiedback-wall failures have resulted from the inclusion of extraneous criteria than from all other causes combined. In many cases, the criterion that the anchor be made only in rock resulted in steep tiebacks, which caused the wall to fail by settling. It is important that critical criteria be met, but too often the full significance of the criteria are not understood.

American Public-Work Procedures

The normal American contracting procedure for public-works projects must be discussed, because it discourages innovation. I do not believe any innovation can flourish in an atmosphere of adversary relationships, which is unfortunately too characteristic of many public projects. The normal practice is for an engineering organization to prepare complete detailed plans of the project. Then bids are solicited. Any contractor who can get a bond can bid.

The system is designed to avoid favoritism and maximize competition. To do this, it is essential that the designs use only well-established methods. The engineering organization's designer must be able to draw detailed plans and write detailed specifications so that the contractor knows precisely what he is agreeing to do. The designer must avoid proprietary products. Since the tieback industry is largely an industry of specialists who have succeeded by developing proprietary products, the designer can hardly find any effective, efficient method for which he can write a method specification. The obvious solution is to use performance specifications, but to many American public agencies this is not a reasonable solution. Both Germany and France have chosen to adopt national code provisions for tiebacks which are performance-type codes. Surely it is possible. Many American public agencies accomplish it simply by making the contractor responsible for the temporary work.

Other public agencies create problems by insisting that their consulting engineers prepare designs for the temporary work. If their engineers are willing to select the proper tiebacks required for the work and are competent to make the selection, then they must consider the fine art of specification writing. If the specification can be written in such a way as to assure use of one of the acceptable tiebacks by a competent specialist, all may be well.

This does not always happen. Frequently the specification is not proper or complete; so when it is followed by an inexperienced contractor the tiebacks do not develop the capacity. The contractor has little or no incentive to overcome the designer's deficiency, nor does the contractor know the fine adjustments which might have caused the tiebacks to be successful. After all, the variables a specialist may control include tieback length, anchor length, soil in which the anchor is made, making the anchor above or below the water table, anchor diameter, bell diameter, the grout mix, the grout-placing procedure, admixtures in the grout, the tendon design, pumping pressure, rate of making the anchor, drilling fluid used, regrouting procedure, etc., etc. Naturally, the specialist will seek to control the variables to get the best results, but often this takes years of experience on thousands of tiebacks. It is unfair to expect an inexperienced contractor to have this competence.

An even greater problem is created by the design organization (usually large) which rejects the approach of selecting acceptable tiebacks and chooses instead to completely design the tiebacks. One project in Washington, D.C., called for belled tiebacks in wet sand. Apparently the designer had discussed the project with an equipment manufacturer and was told the manufacturer could build a machine to do it. The contractor bought the machine, then did everything either the engineer or manufacturer suggested. Nothing worked, so the project was shut down. It was eventually completed several years late at a cost of three times the original bid.

A much more subtle problem for the competent specialist is the garbled specification. Of course, the specialist could refuse to bid any public work with a defective specification, but this would prevent bidding many tiebacks in the public sector. The alternative is to bid and hope to "work it out." This can be quite risky.

Another subtle problem is the specification slanted toward a competitor. Because most tiebacks are proprietary, buried deep in a specification may be a clause which will cost a great deal of money, but which does not benefit the owner at all. For example, specifying a minimum anchor diameter frequently prevents the use of most pressure-injected anchors, even though they may be the best tieback for the site.

Next there is the problem of practical adaptation during the work. The goal of the contracting procedure used on public works is to secure free and open competition by carefully preparing plans and specifications for the project, then allowing open bidding. The low-bidding contractor must follow the bid documents, and in turn will be paid the bid price. The contractor has a responsibility to do a certain amount of work for a certain price. Whether it works or not is up to the designer. While this is suitable for most purposes, it is not adapted to projects where constant fine tuning or innovative techniques may be required during the work.

Finally there is the problem of balancing open competition with the prudent use of well-established methods. Public owners or their consultants, who want to use tiebacks, are faced with more difficult situations than their counterparts in the private sector. They must make selections and enforce them, while not violating public policy. However, seldom, if ever, have satisfactory results been obtained by using a designed tieback that does not utilize a method of construction common in the area.

Summary of First Five Chapters

This book is written both for the tieback user and the engineer interested in tieback design. The first five chapters include most of the information of particular interest to the user. We have discussed the design of tiebacks and some of the methods commonly used. The advisability of selecting successful methods used in the area of the project cannot be overemphasized, nor can the importance of selecting firms experienced with that method. The design discussion should allow the reader to understand how and why the method works, and review the plans. The discussion on testing is quite detailed, because a good test is essential for a performance specification, which I strongly advocate at this time in the evolution of tieback technology.

TIEBACK MATERIALS

Brief discussions of tieback materials were included in earlier chapters. This chapter is a general discussion of the materials commonly used in tieback construction and their merits. Tendons, anchors, connections, centering devices, bond breakers, regrout pipes, and other materials will be discussed in some detail. Materials solely for corrosion-protection purposes are discussed in the next chapter. Selection of the procedure for making the tiebacks may determine some of the materials that have to be used in construction, and instances of this are discussed.

Tendons

Most tiebacks use prestressing steel for the tendon. Our company has not used any other material in years, nor have we seen any project designed by other engineers which did. We use these bars and strands at the same working stress as the codes allow for post-tensioned prestressed concrete. The steel has performed well. Continual high stress over long periods of time is the characteristic use of prestressing steel, just as it is of tieback tendons.

For some systems, individual bars are required. Groups of strands are better for others. Both types of tendon materials will be discussed. Occasionally, button-headed wires or groups of bars are used, but they have limitations which preclude their frequent use, and we do not recommend using them.

The threaded bars we use are alloy steel conforming to ASTM A-722, "Specifications for Uncoated High-Strength Steel Bar for Prestressed Concrete." Two grades of bar are now available in America. The normal bar has an ultimate strength of 150,000 lb/in², and a higher capacity bar with 160,000 lb/in² ultimate strength is sometimes available. In Europe, another bar is available which is approximately 10 percent stronger. The readily available sizes of bar are 1-, $1\frac{1}{4}$-, and $1\frac{3}{8}$-inch diameters. They are sold with a strength guaranteed to exceed the rated ultimate strength. A sample stress-strain curve for a bar is shown in Figure 6-1. For these bars, when used in prestressing, the test and design loads are related to the ultimate load by ACI 318-77. The test load is limited to 80 percent of the ultimate. The design load is 60 percent of the ultimate. As a result, the prestressing design load of 1-inch bars is 38 tons, of $1\frac{1}{4}$-inch bars is 56 tons, and of $1\frac{3}{8}$-inch bars is 71 tons in the steel grade normally available. We use these same design loads for the bars when they are used as tendons for tiebacks.

The test load, 80 percent of ultimate bar strength, should not be exceeded. This means the maximum test load on 1-inch bars is 51 tons, on $1\frac{1}{4}$-inch bars is 75 tons, and on $1\frac{3}{8}$-inch bars is 95 tons. We have tested tiebacks to 90 percent of the ultimate bar strength, but prefer not to. If the bar snaps, it can injure someone. The bar ceases to behave elastically somewhere between 80 and 90 percent of ultimate, so, if we

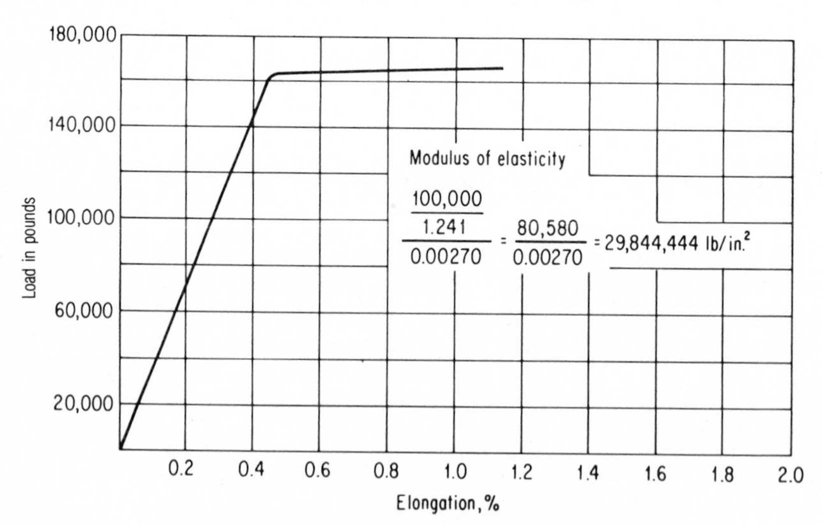

FIG. 6-1 Typical stress-strain curve for prestressing steel.

make a few small errors, we may exceed the elastic range of the bar, which will of course distort the test results. It is just better not to test tiebacks above the test load recommended for the bar. We also prefer to use the design load recommended. Since the relationship recommended by this code is a test load of 133 percent of the design load, we have used performance tests to 133 percent of design load and proof tests to 120 percent of design load. No tieback has failed after passing our proof test, so we do not believe tests to higher overloads are justified.

Larger bars must be used or the tieback design load reduced if test loads to a higher percentage of design load are specified. The highest test load recommended is 80 percent of ultimate strength. If a 200 percent test load is called for, then the highest design load which can be used on a particular bar is 40 percent of ultimate strength. This means the design load of a $1\frac{1}{4}$-inch bar is reduced from the 56 tons we normally use to 37 tons. If 56 tons are required, no bars are normally available with large enough capacities for the required test overloads. The structure may also be overloaded by a test to 200 percent of design. Testing the tiebacks to excessive overloads usually requires more tiebacks for the project and can overstress the structure and cause excessive displacements. You can't improve on no failures, so this is too high a price to pay for "added assurance."

The bars are manufactured and sold in 60-foot lengths. They can be obtained in lengths up to 90 feet, but at a great premium, both in price and time. As a result, tiebacks using bars are usually limited to slightly less than 60 feet long. Couplers are available for the bars, but they are not compatible with the usual installation procedures. When longer tiebacks are required, strand is usually used.

Groups of strands are also used for tendons. Since the tendon capacity can be increased by adding more strands, strands allow larger tendon loads and closer matching of tendon capacity to the design load. The commonly used strand is a $\frac{1}{2}$-inch-diameter strand conforming to ASTM A-416, "Specifications for Uncoated Seven-Wire Stress-Relieved Strand for Prestressed Concrete," and having an ultimate strength of 270,000 lb/in^2. The ultimate strength of a strand is 41,300 pounds. The recommended test load of 80 percent of ultimate and design load of 60 percent of ultimate give a test load per strand of 33,040 pounds and a design load of 25,800 pounds per strand. Several strands have to be combined to make a tendon.

When this strand is furnished for tiebacks, it is usually coated with a corrosion-resisting grease, then a plastic sheath. The grease and sheath are excellent in the unbonded length, but they must be removed in the anchor length if the strand is anchored by bond. The strand manufacturer frequently prefabricates the tendons. This involves cutting, desheathing, and degreasing the individual strands, then assembling them into a unit with spacers, ties, and other hardware.

Strands are cut off a long roll and can be furnished in any desired length without splicing. For some types of tiebacks this makes the strand better than the bars, which are usually furnished in 60-foot lengths.

Both the bar and the strand are high-alloy steel which can be easily damaged. They should not be struck by a welding torch. They should not be roughly handled. They can be fatally flawed by a small dent. They should be treated as what they are—a high-priced special steel for special purposes.

The particular tendon must be compatible with the procedure for making the tieback. In earlier chapters, we stressed the point that the choice of bar or strand is probably dependent on the tieback system used. That should be done by the contractor, but the tendon capacity can be reviewed on the basis of accepted standards.

With some tieback procedures it is necessary to provide centering devices. These are used to keep the tendon near the center of the anchor and are particularly needed in tiebacks where the tendon is placed in grout without using pressure. Experience has shown that, when the tendon has very little grout or concrete cover, it will develop little or no bond strength. The cement protects the tendon while it is in the ground; if it has little or no grout or concrete cover, it has little or no protection. From the above, it is clear the tendon should be centered for both structural and corrosion reasons. Figure 6-2 shows some centering devices. The plastic device on the left is used for centering bars. It has enough strength to carry some bar weight. The right picture shows a centering device on a bar tendon for belled tiebacks. This centering chair is designed for an 18-inch shaft and has to carry both the bar and a heavy end plate. It seems as though a special centering device evolves for each tieback method and even for different contractors using the same method.

The tendons must be connected to the anchor. Excellent bond to grout or concrete is one of the advantages of the deformed bar and the strand. The deformed bar can be placed in the grout or concrete without any special preparation. The bond depends on the grout properties too, but less than 5 feet of embedment will usually be sufficient to transfer the full tensile design load of the bar to the grout in bond. The bond stress exceeds 30,000 pounds per foot on the $1\frac{1}{4}$-inch bar with our pressure-injected tieback. Pullout tests on the bars in grout have been reported with grout-bar bond stresses of 8000 lb/in^2.

The strand can also be bonded to the anchor, but some preparation

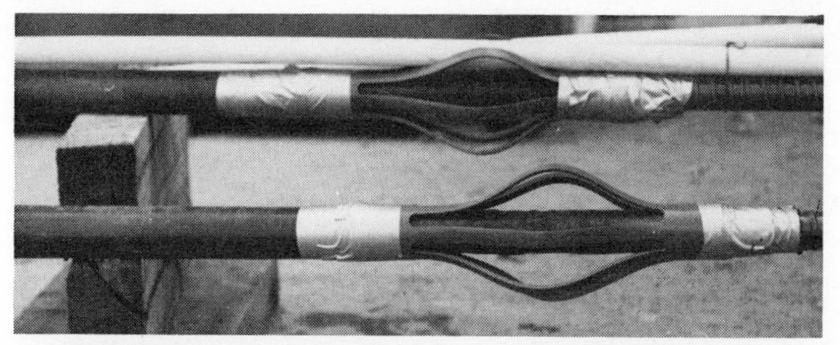

FIG. 6-2a Centering device for bar in small-diameter anchor.

FIG. 6-2b Centering device for bar in belled anchor.

must be done first. The strand should be desheathed and degreased in the bond length. This may involve placing it in solvent baths, steam baths, etc. Clearly, no plastic cover or grease coating can be allowed to remain on the length we wish to bond to the anchor. The PCI code allows a bond between this strand and grout in the range of 4000 lb/in^2. The bond can also be increased mechanically by bundling the bond length, as shown in Figure 6-3.

Plates at the end of the tendon are used with some tiebacks. With our belled tiebacks, for instance, we use a combination of bond and a plate. The plate is held on the bar with a nut and slid to the end of the hole. About 5 feet of bar in front of the plate is also allowed to bond to the bell. The plate is buried in the bell and sized for the entire tieback load.

Compression tubes are another way of securing the tendon to the end of the anchor. They are used in small-diameter anchors, where a plate would overstress the anchor grout. Figure 6-4 shows the arrange-

FIG. 6-3 Spreading strand to increase load.

ment of a compression tube. Compression tubes are used to keep the anchor body in compression. Normally the anchor is in tension, and cracks or discing develop. If the tendon load is applied at the deep end of the anchor, then the anchor will be in compression and no cracks or discing will appear. This is of particular interest for tiebacks which count on the cement as a protection for the steel tendon. The tube is a machined pipe, which can be slid over the end of the bar tendon, then screwed onto the bar. The bondbreaker can extend to the connection between the bar and compression tube. The space remaining can also be filled with grease or cement.

We use points on several of our tiebacks, but they are not part of the load transfer system. The contractor should choose these to suit his technique.

Bond between the tendon and anchor is the usual way of transferring the load, and both bar and strand are excellent for this purpose. In a few cases, it is desirable to secure the tendon to the anchor with plates or a compression tube. The particular load transfer mechanism used may be selected by the contractor for the particular tieback system he is using. In most cases, it is simply bond between the tendon and anchor.

Tendon-Structure Connections

The tendon-structure connection is usually made with standard prestressing hardware. Nuts and plates are used with the bar. Wedges and plates are used with the strand. The hardware for the bar is specially designed so that it can automatically align itself with the bar and keep the thrust axial. This is done by using hemispheric mating shapes on

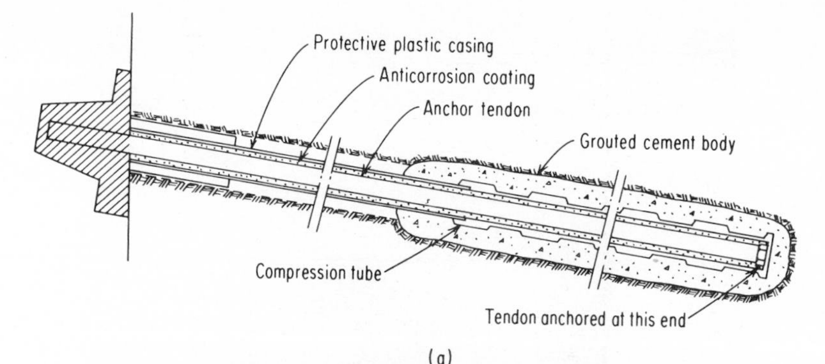

FIG. 6-4a Typical compression-tube anchor.

FIG. 6-4b Section cut through tieback anchor using compression tube.

the nut and plate which can adjust to about a 5° difference. For larger differences special dish-shaped plates and wedges are available. Many of the plates are cast with a bevel to allow a predetermined angular difference between the tiebacks and the structure. One of these is shown in Figure 6-5.

The strand connection is usually made with a special plate and wedges. The strand passes through holes in the plate. The holes are machined to receive special wedges which grip the strand. The wedges shown in Figure 6-6 are pushed into the holes while the strands are tensioned with a jack. When the jack load is released, the wedges grip the strand and maintain the prestress. These plates and wedges are usually furnished by the strand supplier and meet ACI 318.

FIG. 6-5 Bar connection.

FIG. 6-6 Strand connection.

The connection must be made to a point on the structure which is designed for it. In this book, I will not go into this except to stress that it must be designed for the test load. This is frequently overlooked, and, when the tieback is tested, the structure deflects many inches or fails locally.

The connection at the structure may also be recessed. On a dam, for instance, it may be desirable to cut a pocket, place all of the end connection in it, and then patch the hole to match the original surface.

Special end connections for the bars are available, which are designed to be poured into the concrete. These will not allow the bar to be stressed from that end, but instead are designed to be a dead-end anchor. These can allow a bar to be used in a slab without recessing.

Anchors

We have been discussing tendon materials thus far in the chapter. The bars and strands are furnished to ASTM grades of steel by various manufacturers. The end hardware must meet ACI 318 requirements. Next we will consider the anchor materials. Cement grout, sand-cement grout, concrete, and epoxy are the common anchor materials. Their particular properties may be related to tieback construction methods.

Many tiebacks use a neat cement grout. This is a mixture of water and cement. For the pressure-injected tieback, the ratio is unimportant. Most of the water will be driven from the mixture in the ground, and ease of pumping and handling require a wet mix. In contrast, when tiebacks are made in rock without pressure, the mix should have a low water-cement ratio so that it will be strong and not shrink. In this case, it is desirable to use as stiff a mix as can be pumped into the hole.

Grout samples can be obtained and tested following ASTM C109. This procedure should only be followed where the grout is used to make an anchor without pressure or regrouting. The strength of the cubes will be related to the water-cement ratio of the grout mix which usually ranges from 0.4 to 0.45. Techniques which drive water off into the ground will produce grouts in place with much better strength than the cubes indicate.

The cement grout uses regular cement, Type I, or "high-early" Type III, meeting ASTM C150 requirements. We have made grout samples with Type I and tested them to more than 20,000 lb/in^2. Type III cement is a finer-ground cement. When used in concrete, the concrete reaches its design strength in 7 instead of 28 days. This is not readily available, and the same result can be obtained by using more regular cement in the concrete mix; so normally we use Type I cement.

We mix the grout in two ways. Paddle mixers combine the cement and water into a uniform mix for pumping. Shear mixers use much more power in mixing to produce a colloidal or more stable mixture. The shear-mixed grout is a smoother mixture, which is essential for some operations. For example, when a regrouting procedure using packers, valves, and small holes is involved, the grout must be mixed in this way.

The first grouting of holes to be regrouted may have a special grout mixture, which is harder than the ground but weaker than normal grout. This is usually accomplished by adding bentonite to the grout. We have run tests with the cement-bentonite-water grouts which show that the bentonite ratio can be increased until the strength approaches zero. The purpose of this special grout is to permit the grout to fracture around the regrout tube. Grouts of this type have been developed by specialist organizations as part of proprietary systems.

We have used commercially available admixtures to accelerate the hardening of the grout in special cases. With these accelerators we have been able to test tiebacks the day after they were grouted. This is the only admixture we now use in cement-water grouts.

We particularly avoid admixtures designed to cause the grout to swell when it sets. These admixtures usually act by causing gas to be released in the grout. They weaken the grout, and the gas they release is also capable of damaging the tendons by accelerating stress corrosion. No one knows enough about stress corrosion, but it is foolish to create a condition favorable to it.

Another grout we use is a sand-cement-water mixture. This is used with the hollow-stem-auger tieback in particular. The grout is purchased from ready-mix suppliers in mixer trucks and pumped by a regular concrete pump. The water-cement ratio is maintained at about 6 gallons per sack of cement. We normally add some fly ash to the mixture to improve its pumpability. The strength of this grout can be checked with regular concrete cylinders.

For belled tiebacks we use transit-mixed concrete. The concrete should be a 3000- to 4000-lb/in^2 mix.

Another grout we use is an epoxy resin, which is sold under several trade names. This is a special product made for anchoring rock bolts in rock. The material is furnished in sausages. Each sausage contains the resin and the catalyst in separate compartments in a casing about an inch in diameter and a foot long. Setting time can be varied by the manufacturer, so that a choice of times is available when buying the material. We usually continue twisting the bar in the hole to stir the materials together until the material sets. About 5 minutes is a reasonable time.

Bond Breakers

Bond breakers are desirable on most tiebacks. They permit the bar to be surrounded by grout or earth but not bonded to it. A polyethylene pipe is usually used for this purpose. It is just large enough to be a slip fit on the bar. To prevent grout from entering, it is usually secured at both ends with tape. This can be standard material available at any plumbing wholesaler, though we buy it to our specifications. The grease and sheath furnished on strand is designed by the manufacturer as a bond breaker and is satisfactory for that purpose.

Regrouting

Regrouting pipes and tools are usually specialty items developed for a particular system. They are furnished by the specialist who is doing the work. They are of two basic types, which I will call the first open system and the packer system. In the first case, a regrout pipe is inserted into the first grouting of the hole. The pipe has a number of holes in it, which are protected by one-way valves. When grout is later pumped into the pipe, it must force a valve to open by breaking the grout surrounding the valve. This may require hundreds of pounds of pressure. Our experience has been that frequently the grout cannot be broken, so no regrouting is possible. I am convinced that in most cases, as soon as one valve opens, the pressure drops and no more can be opened. Therefore, all grout pumped goes through the same valve, the first one to open. For this reason, several pipes of this type are often used in the same anchor.

The packer system also uses a regrout tube with valves, but it is designed to use a packer, so that each valve can be isolated and the grout pumped only through it. Figure 6-7 is a picture of one regrout system we use, showing the pipe with holes and then with the rubber valves in place over the holes. A small pipe with packers slides into the regrout tube. Water expands the packer against the tube above and below the valve. When grout is pumped down the pipe, it enters the space between the packers and pressure is increased until it begins to exit through the holes and expand under the rubber valve. When it applies pressure to enough area of the valve, it breaks the grout. It then flows out and against the soil. When enough grout has been pumped,

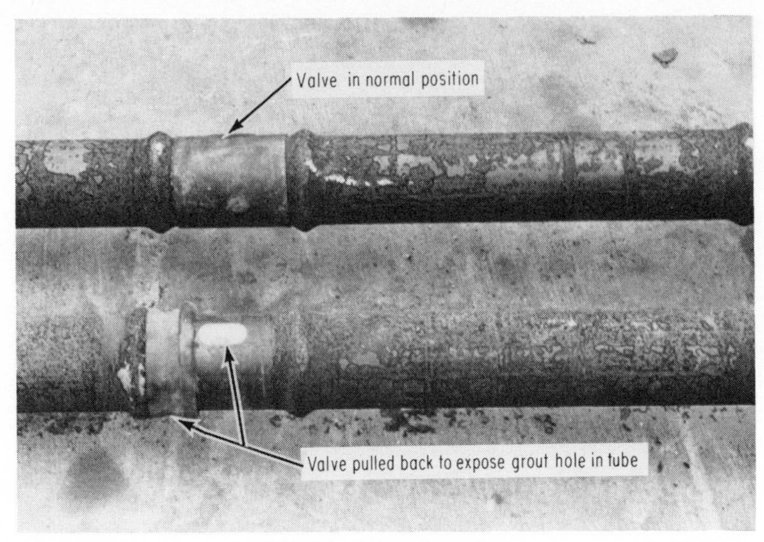

FIG. 6-7 Regrouting valve.

the pressure is released, the valve closes back over the hole, and the pressure of the grout against the soil is maintained. The packer pressure is released, and the process can be repeated at another hole.

In this chapter we have discussed materials for tiebacks. Through the years, other materials have been tried and evaluated, and certainly new ones will prove desirable in the future. It would be impossible to describe all of the materials used with the various proprietary systems. In most cases the important materials are ones all foundation engineers are familiar with—steel, cement, and water—used in an effective way.

7

CORROSION PROTECTION

When we design tiebacks as a part of permanent structures, so that they have to last as long as the structures, we must rely on good engineering judgment. The oldest corrosion-protected tieback I know of has been in service for only 40 years, and most structures are designed to last longer than 40 years. We have to recognize that we are dealing with a new technology which has great apparent advantages, but cannot have its long-term efficacy proven by citing examples. We can cite some similarities to other types of projects, but we are ultimately forced to make an engineering evaluation of the resistance of the tiebacks to corrosion. Corrosion mechanisms, corrosion rates, the effects of the earth on the corrosion, and methods of preventing corrosion by coatings or encapsulation are the topics which we have to study. Reasonable judgments are being made of these factors, and tiebacks are being used as vital parts of permanent structures. Figure 7-1 is a photograph of the cable-suspended roof of the Olympic Stadium in Munich, which is one of the most spectacular uses of this technique. The ends of the roof cables were anchored with tiebacks in sand.

In this chapter I shall consider the problem of the gradual loss of tieback capacity through deterioration of the tieback materials. The next chapter considers creep, the gradual loss of tieback capacity caused by slow anchor movements. These two problems are the most difficult to evaluate in the design. We believe that they can be intelligently studied and reasonable engineering judgments made. Only time

FIG. 7-1 Tiebacks support Olympic Stadium roof.

and experience can verify the accuracy of these judgments, but there are no other problems which should prevent the use of tiebacks in "permanent" structures.

Evaluating Corrosion

Tendons are steel, and steel corrodes under certain conditions. The anchors are dependent on cement for their strength, and cement in the earth has been attacked and destroyed. We have to evaluate the probability of either occurring and design to prevent it if we are to use these tiebacks. The problem can be separated into two parts: the probability of the tieback being damaged by deterioration of the steel or cement and methods to prevent it.

Tiebacks for short-term use seldom have any special protection. Our company, on several occasions, has had "temporary" pressure-injected tiebacks in service for more than 5 years. No problems developed. We have discussed this with European and Japanese experts, and they have had the same experience. The tiebacks had bar tendons, pressure-injected anchors, and polyethylene bond breakers covered by grout. The head connection was a plate and nut with no corrosion protection. The tiebacks were in virgin soil and a generally noncorrosive environment. We have also excavated and exposed some similar tiebacks and found no apparent corrosion. From these experiences, it appears some tiebacks may not need any special corrosion-resisting protection.

There are also some places we will not install tiebacks because of corrosion fear, failing bulkheads in saltwater harbors, for example. If the steel in the particular environment is being rapidly corroded, clearly the tieback steel is in danger. If tiebacks are to be used, they must have a level of protection in such an environment capable of preventing corrosion. The rest of this chapter is devoted to evaluating the corrosion and designing the corrosion-protection system.

One type of steel corrosion is known as surface corrosion—rusting. As its name implies, it occurs over the surface of the steel, though it is often uneven. It occurs over a period of time and is usually considered time-dependent. The usual design practice for structures in soil is to study the rate of corrosion and, from this, calculate the amount of corrosion that will occur over the life of the structure. Then the decision is made either to make the member oversized to allow the rusting or to protect the member from this type of corrosion. The threat of corrosion may be greatest at one part of the tieback, for instance at the connection to the structure, so the level of protection in the anchor length, unbonded length, and head connection will each have to be considered separately.

Stress corrosion is more likely to cause failure of tieback tendons than rusting. The exact causes of stress corrosion are poorly understood, but it causes a localized crack in certain high-strength steels. Certain steels (heat-treated and quenched, for instance) may be more susceptible. Most engineers feel stress corrosion will not occur if the environment in which the steel is used is properly controlled.

The cement or concrete of the anchor is often the only tendon protection in the anchor length. I have never heard of any tieback which failed as a result of tendon corrosion in the anchor length, regardless of the type of protection. There are good engineering reasons why this should be so, yet we continue to postulate failure mechanisms and design for them. I will discuss why it is reasonable that the simplest anchors are lasting, before considering the observations which have caused some engineers to recommend more elaborate tendon protection. Surface corrosion is an oxidation of the steel, which does not occur deep in the earth because oxygen is not available. Tests by M. Romanoff (1962) at the National Bureau of Standards on steel piles clearly show this. Romanoff stressed that most earlier corrosion data was obtained on steel located in backfill and that backfilling causes a drastic increase in the oxygen content of the soil, which promotes corrosion. When he found no appreciable corrosion of steel pilings that were driven deep into the original ground, he attributed it to lack of oxygen. He reports this to be true regardless of soil properties. If this research can be extended to tiebacks, the tendon does not need any corrosion protection in the anchor zone. Yet, almost all tendons receive near-perfect protection from the grout in which they are anchored. For example, a study made of methods to protect post-tensioning steel from corrosion when used in nuclear containment vessels found that a grout coating provided complete protection to the steel. This study also concluded that with portland cement crack widths up to 0.03 inch did not result in loss of protection. (Griess, 1975) We know discing or cracking of the tieback anchor body must occur because the grout cannot strain in tension as much as the steel does at its working load. Cracks have also been observed in anchors which have been dug up. Figure 7-2 shows one such test site. From observations such as this, we know that the cracks in the pressure-injected anchors in the area of maximum tendon strain are spaced about 1½ inches apart. If no strain occurs in the anchor grout, so all strain is represented by crack width, then the crack width should be equal to the bar strain at the tendon load. On this basis, the crack width would be 0.005 inch. This is not a wide enough crack to prevent the grout from completely protecting the steel. On this basis, no special corrosion protection is necessary for the tendon bonded to a grout anchor.

Protecting Tendons by Encapsulation

Some engineers still feel more protection is required. Frequently, anchors in rock are made by grouting the anchor length, then redrilling the hole and pressure testing it to be sure the first grouting is watertight. This prevents flowing groundwater from carrying dissolved oxygen to the tendon. The cracking has caused other engineers either to require the tendon to be protected by more than the grout body or else to redesign the tiebacks so that the grout body is put in compression, not tension, and thus eliminate the cracks. I am familiar with a number of these techniques, all of which are based on covering the tendon and most of which are patented. We have developed a system which protects a bar tendon with an electrostatically applied epoxy. The integrity of the coating can be checked electrically, it can be easily patched, and it is able to transfer the bond between the steel and grout without being damaged.

FIG. 7-2 Excavation to observe tieback anchor: (*a*) row of tiebacks; (*b*) closer view of one tieback.

Another system, developed by Karl Bauer and also being sold by the bar manufacturer, places the bar in a corrugated plastic tube, then fills the annular space with grout, inserts the assembly into the tieback hole, and grouts it into position. The assembly is usually made in the shop and shipped to the jobsite. With this procedure, the bar is protected by first grout, then the plastic tube, then more grout. Bachy has developed another system, which we prefer to use with strand tendons.

After the hole is drilled, a tube is grouted into place. The tube has valves which, together with a special grout pipe, allow grout to be selectively pumped into the ground and thus increase the capacity of the anchor. Then the strands are inserted into the tube and grouted into place. Thus the tendon is protected by grout, tube, and grout again. The last method is intended to transfer the load to the grout body in such a way as to cause it to be in compression. This method, which uses a compression tube, was discussed in an earlier chapter. In general, the tendon is covered for most of its length with a plastic tube and coal-tar epoxy, but at the end it is screwed into a steel tube. The bar cannot bond to the anchor, but depends on the steel tube to transfer the load by bond and compression. In this way, the bar is protected by a coating, a steel tube, and also the grout body, which cannot crack because it is in compression.

All of these systems basically depend on a coating to provide additional corrosion protection to the bar. As far as I know, the need for this coating is based solely on the presence of the cracks, and no research has been done anywhere in the world relating the cracks to the corrosion of tieback tendons. Since no tiebacks have failed as a result of tendon corrosion in the anchor length, we are forced to evaluate the need for, and level of, corrosion protection on the basis of studies made for other purposes. In the years to come, more research specifically considering tiebacks will have to be done. Until then, we are left with engineering judgment, which so far has a perfect record of no tendon corrosion in the anchor length.

Anchor Deterioration

We must also be sure the grout in the anchor length will not deteriorate with time. Since the anchor body is made of portland-cement grout, or concrete, we must be sure that neither the soil in which we anchor nor the groundwater will attack it. As with the corrosion of the tendon, we have a great body of engineering information available concerning deterioration of grout or concrete, but nothing specifically has been done on tieback anchors. If we analyze the available data, we find most anchors are in no danger from either soil or the groundwater. The soils which attack cement are high in natural sulfates, and Type II cement can be used, since it is specifically formulated to resist sulfate attack. The groundwater can carry with it an almost infinite variety of aggressive agents which may attack the grout, sea water, for example. Chemicals may have leached into the ground, or acid from mine wastes. If there is any reason to suspect such problems, the most prudent course is to obtain samples of the groundwater and have an analysis made. Then a special grout may be selected to resist the attack. Usually this simply means choosing the appropriate cement. When we use any cement other than ordinary portland cement, we should also be sure that it will continue to be a protection for the tendon steel. There is not often a danger of deterioration of the cement of the anchor body, but, when it is suspected, we can analyze the soil and groundwater and select a cement which will resist deterioration. I know of no case where tiebacks could not be properly designed to assure long-term dependability of the cement selected for the anchor body.

In the last paragraph we discussed the method of designing the anchor to assure that it was not attacked and deteriorated by either the soil or groundwater. In the discussion that preceded that, we discussed corrosion of the tendon and ways being used to prevent tendon corrosion in the anchor length. Though these designs must extrapolate from other studies on steel corrosion and cement deterioration, we can use them with a reasonable degree of confidence for anchor design.

Corrosion Potential in the Unbonded Length

The corrosion protection of the tendon in its unbonded length must also be considered. In general, this must be considered a more aggressive environment than the anchor length, since the tendon often emerges from the soil into the atmosphere. The usual way of protecting the unbonded length is, therefore, to coat the steel with a dependable coating. A bond breaker is used over the protective coating so that no load will be transferred to the soil in this length. The only other commonly used procedure is to grout the anchor but not the unbonded length, test and stress the tendon after the anchor has cured, and then grout the unbonded length. I definitely prefer the method using a coating and bond breaker, but will discuss both methods in more detail.

An elastic adhesive and grease are the common coatings used. Since the tendons are usually bars or strands used for post-tensioning applications, we again rely on studies done to assure corrosion-resisting coatings in post-tensioned applications. Either coating has been found to be satisfactory. Generally, the elastic adhesive is used on bars, and grease is used on strand, because it is easier that way. We have to coat the bars ourselves, and the method we use is both effective and economical. We slip a special plastic tube, which is coated on the inside with the adhesive, onto the portion of the bar we want to protect. We then heat the tube, which shrinks onto the bar, squeezing the hot, tacky adhesive into intimate contact with the deformations of the bar. Then we slip another plastic pipe, which is just large enough to allow a slip fit, over the first pipe and seal both ends with tape. This second pipe prevents grout from encasing and bonding to the shrunken pipe, which is no longer smooth but has assumed the rough shape of the threaded bar. When the bar and its protective coatings are installed in a tieback, this length is protected from corrosion and also isolated from bonding to grout which may surround the outer pipe.

When we use strand, the manufacturer supplies it with a grease protective cover under a plastic pipe. The grease is a specially formulated grease with excellent corrosion-resisting properties. The grease is first applied to the strand and then the pipe formed over it. This plastic pipe serves as a bond breaker and protects the grease from damage by rubbing, grout contact, or other causes. The greased and sheathed strand can be grouted into a tieback and will be both protected from corrosion and free to strain elastically in the unbonded length.

The third method which we have used is not as satisfactory. In general, with this method the anchor length is grouted in such a way that no grout is left around the strands or bar in the unbonded length. This is accomplished by injecting into the hole only the amount of grout calculated to make the anchor. Alternatively, the entire tieback length is grouted, and then the grout in the unbonded zone is flushed away. After the anchor has cured, the tieback is tested and locked off. Because at this stage the tendon is in an open hole, it of course is unbonded. The usual practice is to test and lock-off the tendon, then grout the open hole. Occasionally a pipe is inserted around the tendon in the open hole, so that grout can be pumped between it and the soil and also inside it around the tendon. In addition to being more expensive, I don't believe the protection is as good. The joint between the two pours has to be particularly poor grout no matter which method we use.

In the preceding discussion we have discussed three methods commonly used to protect the tendon from corrosion in its unbonded length. All three methods are based on covering the steel with coatings which have proven effective for protecting post-tensioning steel, namely, elastic adhesives, grease, and portland-cement grout. Earlier in the chapter we discussed some special methods of protecting the tendon from corrosion in the anchor length, and additional protection of the unbonded length is also a part of these patented systems. For example, we coat the entire bar with the electrostatically applied epoxy, so that under the elastic adhesive is another layer of protection. The bar is covered with a smooth plastic bond breaker in the Bauer and compression-tube systems, under which a grease or epoxy is injected. In the Bachy system, a smooth tube connects the grouted anchor pipe to the surface, and it is filled with grout after the tieback is tested and locked off. As a result, the bar is completely encapsulated in a coating for all four proprietary systems. The tendon can also be tested in any system without transmitting significant load throughout its unbonded length. I feel we can design a corrosion-protection system for the unbonded length of the tendon which will be reliable.

Protecting the Connection to the Structures

I am less confident about designing a connection between the tieback and the structure which will not corrode. Tiebacks have been successfully used in some of the most corrosive environments imaginable, such as the bulkhead in a salt water harbor shown in Figure 7-3. This is an example of the successful design of the end-connection protection in an aggressive, hostile environment. Many other examples could be shown to prove it is feasible. Once again, we are forced to design the connection based primarily on facts developed for the design of post-tensioning end connections. The key elements of the end-connection design are assuring that the corrosion protection of the unbonded length is continuous into the end connection and that the bars, strands, wedges, nuts, or plates are adequately protected. Of course, since the tieback must be stressed, it is essential that the necessary movement of the tendon be allowed. Most end connections designed to resist corrosion combine a "trumpet" and a "cap." The trumpet is designed to overlap the unbonded tendon length, allow movement, and protect

FIG. 7-3 Tied-back wall in salt water.

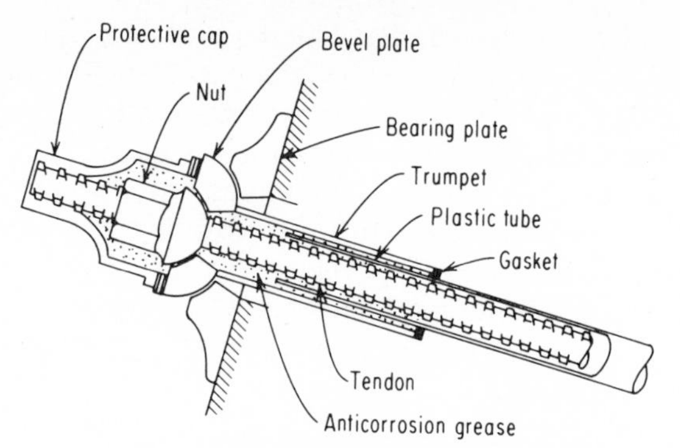

FIG. 7-4 Protection of the head of a tieback.

the strand or bar where the unbonded length covering ends. The trumpet is connected to the bearing plate. The cap is designed to protect the nuts or wedges and the end of the tendon which passes through the bearing plate to be gripped by them. The cap can be a protective material, such as grout, in which the nuts or wedges are embedded, or it can be a shaped plastic or steel cap which is then filled with grease or grout. A schematic illustration of one protection for a tieback head is shown in Figure 7-4. Note that the trumpet and cap are both secured to the plate, so that the trumpet and cap combine to form a closed chamber in which the grease is confined.

The trumpet can be of steel or plastic pipe. It should be secured to the bottom of the bearing plate with a threaded or other tight joint. The gasket at the end is used to permit a slip fit between the trumpet and bond breaker during stressing. It, of course, must be in place before the tieback is tested. Then after the tieback is successfully tested, the cap is screwed on and filled. We have used both steel and plastic for this type of cap.

Probably any cap which has been used to successfully protect post-tensioning tendons in similar conditions can be used for this application. All post-tensioning suppliers make a recommendation for protecting their tendons, which is often to use a cap. Of course, they sell the caps they recommend, so it makes them readily available.

The level of head protection required is very uncertain, and a great deal of judgment must be exercised. Earlier in this chaper I said that we have had hundreds of tiebacks with no corrosion protection in service for years and have not had any fail from corrosion at the head. But we try to analyze the cause of failure of any tiebacks we know of and have concluded that corrosion at the head is one of the chief causes. So, head protection is critical.

The simplest form of head protection is to spray the nut, tendon, and plate with a grease and antirusting paint. In some environments, this may well be adequate. Another common practice is to recess the head connection into the concrete structure and fill the cavity after the tendon is tested and stressed. This is apparently a very effective technique. The trumpet and cap shown in Figure 7-4 or some similar arrangement is probably the most common. In addition to protecting the head, this procedure allows us to remove the cap and check or restress the tendon. The tendon can be completely encapsulated by using a plastic tube, such as the Bauer protection discussed earlier, together with a plastic trumpet, cap, and pad beneath the bearing plate. In this way the tendon can be completely isolated from any stray ground currents where they are a consideration. Of course, as more extensive protection becomes necessary, the cost increases, so it is important to evaluate the protection really needed.

The last dozen pages have discussed several corrosion-protection methods. This is a subject which is difficult to reduce to numbers. On the one hand, most of the decisions on the level of protection must be made by extrapolating studies made for other purposes to the case of tiebacks. The tiebacks will be buried in soil, and soil is one of the least consistent materials engineers have to deal with. Also, the provision of some forms of corrosion protection prevents the use of the most effective tiebacks for the particular soil. Finally, most of the frequently used systems have been developed and patented by particular organizations, who use them with particular tieback techniques. The difference in cost between some of the systems is a factor of 2 or 3. A prudent approach is to consider both cost and the protective system before making a choice.

Before leaving the subject of corrosion, we should consider tendon selection and admixtures.

Tendon Steels

We use steel post-tensioning bars and strands as our tendons. For years we used only the bars. The best shape to resist corrosion is a bar since it has the maximum ratio of cross-sectional area to surface area. If surface corrosion, rust, is the phenomenon we must consider, then we must consider some depth of corrosion. A hundredth of an inch around a bar may not reduce the area much, but a hundredth of an inch on each wire in a strand tendon may. So the shape of the tendon is important.

The selection of the tendon steel is, of course, vital. We use only steel approved for post-tensioning applications. Many tiebacks have failed from corrosion as a direct result of choosing the wrong steel for the tendon. In several cases, the tendon was fabricated from steel not approved for use today in post-tensioning. Occasionally, permanent tiebacks have been designed using steels with lower yield points. This, of course, increases the area of the bar, so it makes surface corrosion less significant, but I believe it has usually been done out of concern for stress corrosion. Stress corrosion is one of those phenomena which are greatly feared and poorly understood. The experts are not in agreement about its causes, mechanisms, or cures. It seems to occur only in steels alloyed to achieve high allowable stresses; it seems to cause a crack to propagate toward the center of the steel until it is sufficiently weak to fail; it progresses rapidly; some believe it occurs only under high stress conditions; and it occurs at discrete locations, perhaps where hydrogen ions are present. From this mass of opinion we have to draw some conclusions. Prestressing steel, both bars and strands, is regularly used in critical applications without stress corrosion occurring. So, if we can use it properly, we can overcome the threat of stress corrosion. The manufacturers are, of course, aware of this concern and have controlled the metallurgy to prevent stress corrosion from occurring. Coating the bar and embedding it in cement help protect it. Carefully protecting the head is very important. These are the same precautions which are effective in reducing surface corrosion.

In an earlier chapter I mentioned that certain admixtures designed to cause the grout to expand release hydrogen ions. Since free hydrogen ions have been cited as a condition promoting stress corrosion, we have avoided these admixtures. We have been careful to use portland-cement grouts, which are ideal for protecting the tendon from corrosion. Since this can be done in most cases, we have avoided admixtures. They should not be used unless their action and properties can be checked and found to be compatible with the corrosion protection.

In general, tiebacks are protected from corrosion by encapsulating them. This means they are covered by protective layers to isolate them from contact with the soil, groundwater, or stray electric currents. The anchor length, the unbonded length, and the connection to the structure must all be appropriately protected, and each has been considered separately.

8

MORE ON TESTING

In Chapter 4 a testing routine for tiebacks was recommended. This chapter considers in more detail the physical significance of these routine tests. The tests are usually reported as load/deformation curves, and in this chapter we will discuss the fact that different methods of making tiebacks have different characteristic load/deformation curves. Testing errors, particularly those caused by errors in reading the loads or deformations, also affect the reported results. A more detailed discussion of creep then follows. These tests are used to assure the reliability of every tieback incorporated in the project. The next part of the chapter is devoted to preconstruction testing of tiebacks, which usually is performed at the direction of the project engineer. The chapter ends with a discussion of long-term tests to verify tieback performance.

Characteristic Load/Deformation Curves

The purpose of the proof and performance tests is to verify that the tiebacks will carry the design load. The fact that particular types of tiebacks have characteristic load/deformation plots, which are reasonable, has led to an emphasis on this plot for evaluating tiebacks. We must remember that the test is to verify capacity, and the load/deformation measurements are only made to help do this. Actually, the deformation may have little or no real importance since it is induced by the preloading. The key purpose of the test is to verify that the tieback anchor will carry the load for the life of the structure.

To understand the load/deformation curves, we must consider the tieback design. Usually a portion of the tendon is unbonded, so that all of this length of tendon should be stressed to the particular test load. The theoretical strain can then be calculated based on the steel properties. The tendon must be secured to the anchor, and the test load causes strain of the anchor. For simplicity we refer to anchor strain as the strain at the point on the anchor where the attachment of the tendon begins. Anchor strain is very complex. It is affected by many variables, including the method of making the anchor, the soil, and the method of transferring load from the tendon to the anchor. To understand the load/deformation curves, we must understand how the anchor deforms. We now have enough experience to consider some characteristic deformation patterns.

To illustrate, three successful performance test results are shown in Figures 8-1, 8-2, and 8-3. On each, the theoretical strain of the unbonded test length of the tendon is shown. The first tieback has a

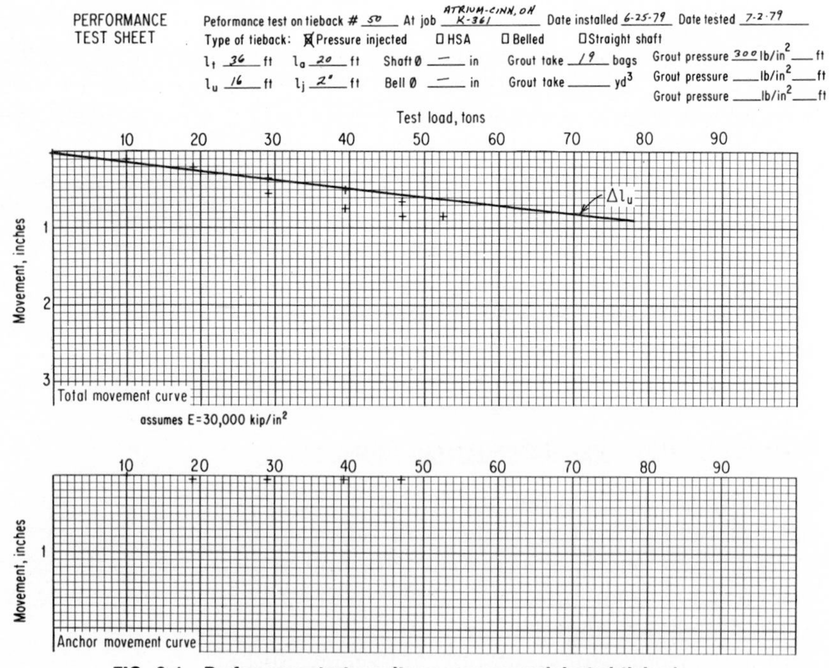

FIG. 8-1 Performance-test results on a pressure-injected tieback.

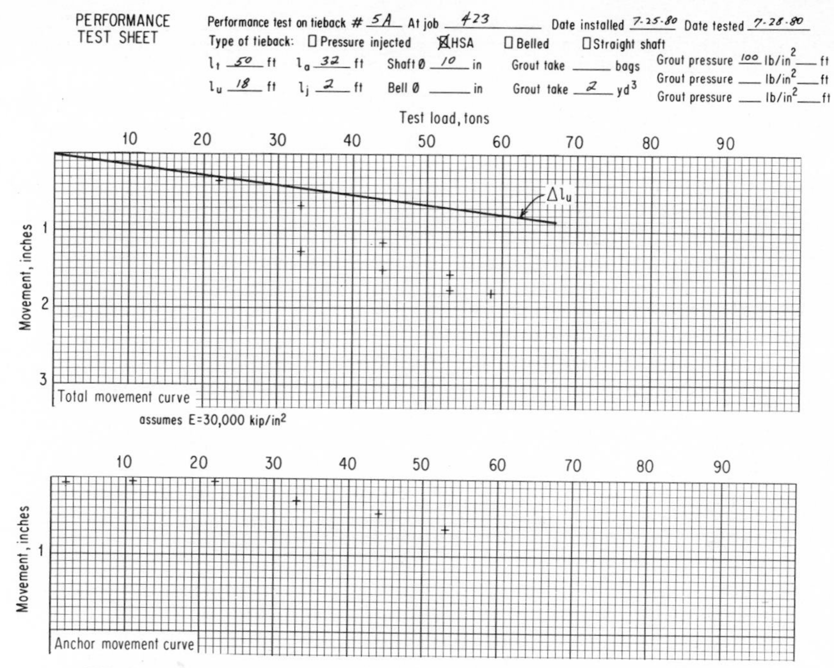

FIG. 8-2 Performance-test results on tieback made with hollow-stem auger.

pressure-injected anchor in dense sand. Note that the residual anchor movement for this tieback is very small. As a result, total tieback movement closely follows the strain in the unbonded length. This fact, which is characteristic of this type of tieback, has led to the adoption in many specifications of an acceptance criterion for this tieback related to the strain of the unbonded length of the tendon. This criterion is not valid for the other two types of anchors, which we see move a great deal more. The second tieback, Figure 8-2, was made with a hollow-stem auger in soft clay. The third tieback, Figure 8-3, is a belled tieback made in disintegrated rock. It is clear that each type of tieback has a characteristic load/deformation plot.

Note that the hollow-stem auger and belled tiebacks elongated two and three times as much under load and, for each, the anchor movement was a major contributor to that elongation. For example, at 53 tons anchor movement of each of them was ten times more than that of the pressure-injected anchor. The jump in anchor movement of the hollow-stem-auger tieback after 22 tons is related to discing, or cracking, of the anchor grout. For low loads, the grout of the body carries some tension, but as strain increases the grout cracks. For pressure-injected tiebacks, this occurs at such low test loads it is not observable.

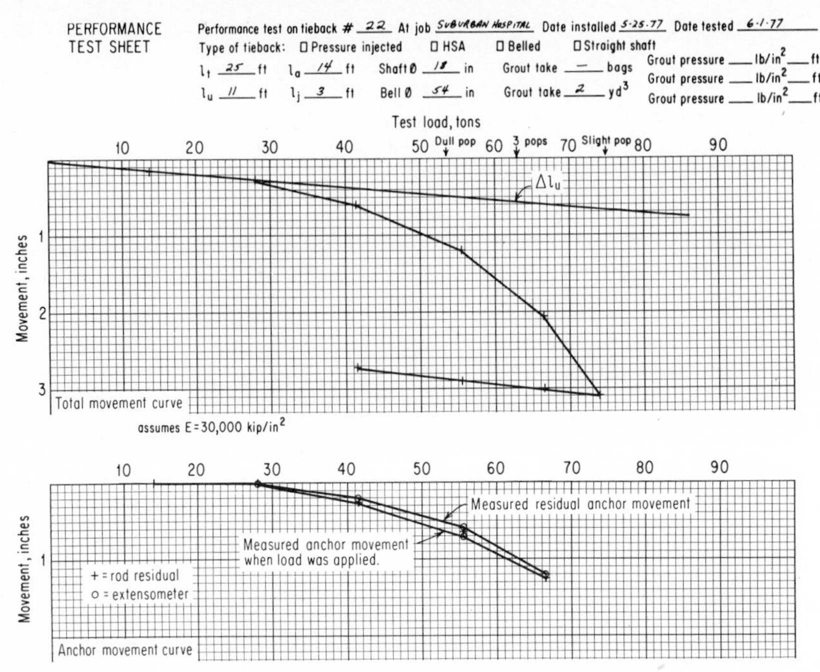

FIG. 8-3 Performance-test results on belled tieback.

For grout bodies with an area of about 100 square inches, such as the hollow-stem auger has, it is significant. When the grout cracks, two things happen. The stress formerly carried by the grout is transferred to the bar, and it strains and thus increases the anchor movement. Also, the load transfer between grout and soil is changed. The increase in relative movement of the disced portion of the anchor should increase the load transferred, but, as additional load is carried by the tendon, it transfers load further back into the anchor. For this type of tieback, which is used in softer soil, larger anchor movements are normal, as is the pattern of abrupt anchor movements resulting from fracturing the anchor grout.

A third example of a characteristic load/deformation curve is the belled tieback. The tieback of Figure 8-3 had a tendon bonded to the anchor. The anchor was a 54-inch-diameter bell, with 8 feet of shaft concrete in front of the bell. As a result, the ultimate capacity of the shaft was first developed by the test, then further load was carried by the bell. This resulted in very small anchor movements until the ultimate load transfer capacity of the shaft was reached between 28 and 42 tons. As the load was increased, the shaft was cracked, but probably carried the same load. All additional load was carried by the bell, which moved much more because the bearing modulus is lower than the shear modulus of the soil. This particular test had an extensometer placed at the front of the shaft so that we could study anchor movement further. Note that there was very little elastic rebound of the anchor and the anchor movement measured with the performance test is in good agreement with the extensometer. It is a characteristic of this type of tieback for the rate of anchor movement to increase abruptly when further load is carried only by the bell.

These three examples are cited only to demonstrate that the load/deformation reports from performance and proof tests are clearly affected by the method of making the tieback and the soil. The acceptance criteria must consider these characteristics.

The three examples also clearly show that using the performance test to quantify anchor movement is reasonable. Note that there is good correlation between total tieback movement and the sum of measured anchor movement and calculated strain of the unbonded tendon.

Our experience has taught us that each of the three tested tiebacks should be accepted on the basis of its ability to carry the test load. We also recognize the patterns of load deformation as being characteristic of the tiebacks. These three tests are cited as examples only to show that the load/deformation results are affected by the tieback design. Tieback designers must be familiar with this fact and learn to recognize characteristic anchor movements. This is an area of tieback testing practice which does not seem to be well-understood and has caused a great deal of unnecessary problems when acceptance criteria appropriate to one tieback method have been applied to others.

In particular, the procedure adopted in the German DIN 4125, which is for pressure-injected tiebacks, has been copied and applied to other methods of making tiebacks. I doubt that very many tiebacks made with bells or multiple bells, as in England, will meet the acceptance criteria for that test. It is not that they are bad tiebacks. Usually it means that some engineer has selected an inappropriate test procedure. The practical effect of rigorously enforcing the test and its acceptability criteria is to favor the tieback method for which it is appropriate and discriminate against all others. The remedy is to learn a little more about testing and recognize that the shape of the load/elongation curve is a function of many variables, including at least anchor shape, soil, and unbonded length.

The characteristic load/deformation curve from a tieback test will be affected by the design of the tieback, as we have seen. One aspect that has not been discussed yet is the appropriate unbonded length. This is the length of the tendon between the test jacks and the point where the anchor is first secured to the tendon. In simple terms, it is the length of tendon which is fully loaded to each of the test loads. Thus, to the unbonded length of the tendon, as it is commonly understood, we have to add the length of tendon needed by the test jacks and other test hardware. We also have to consider that the unbonded and anchor lengths may overlap.

The majority of all tiebacks we make use a tendon which is bonded for the full anchor length. For this type of tieback, the total tieback length can be divided into unbonded tendon length l_u and anchor length l_a. Of course tendon strain takes place in the anchor, but this is affected by many factors, as discussed earlier. For two pressure-injected anchors, shown in Figure 8-4, it is clear that the stress decreases rapidly. In fact, for the upper case, no stress in the deep half of the anchor could be measured.

Some engineers believe tieback capacity can be increased by avoiding the large rate of load transfer at the front of the pressure-injected anchor and placing some of the anchor in compression. This is done by continuing the unbonded length of the tendon into the anchor. The anchor and unbonded tendon lengths overlap. Total tieback movement under the test loads will be increased. Usually the anchor strain is not much affected by this procedure, but the unbonded tendon length is increased. If the tendon is unbonded for the full length of the anchor, but transfers its load to the anchor at the deep end of the tieback, we have a longer unbonded length and should observe more movement when testing. The advantage of this method of making tiebacks is that the anchor is in compression; hence it should not crack.

The unbonded length of the tieback is selected by the designer. It is simply the length of the tendon between the structure and the point at which the tendon is secured to the anchor, which carries the full tieback load. The unbonding is usually accomplished by placing the tieback in a loose-fitting plastic pipe, in which it is free to move without load transfer. The purpose is to assure that tieback load is transferred to the anchor and nowhere else. We also use greased and sheathed strand in the unbonded length of our tiebacks. Obviously in either of these systems some frictional transfer may occur, but it will be a small amount. The third method is to test the tiebacks while the tendon in the unbonded length is in an open hole. Even here some load may be transferred if the hole is not straight and the tendon pushes on the side when it is tensioned. A truly unbonded length is rare. Some friction in this length usually occurs. However, it is small so that in spite of it we can verify that little or no load is transferred in this length.

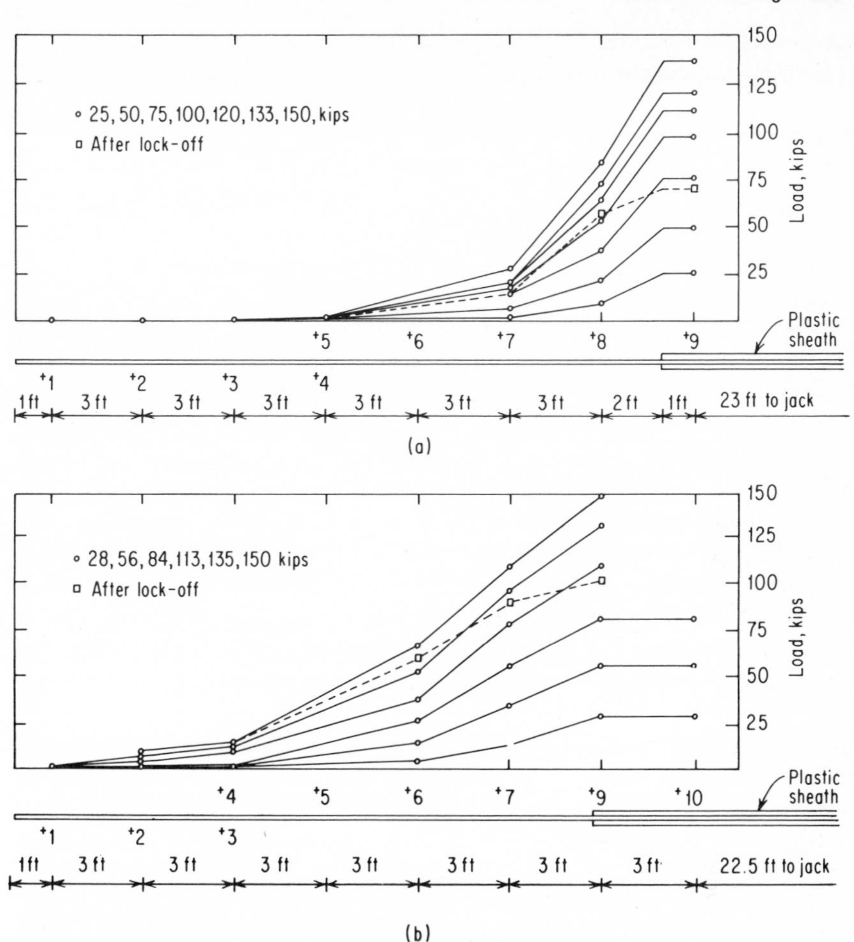

FIG. 8-4 Tendon force in anchor length during testing of pressure-injected tiebacks: (*a*) National Capital Bank; (*b*) 1800 Massachusetts Avenue.

The theoretical strain in the unbonded length Δl_u is:

$$\Delta l_u = \frac{Pl_u}{EA}$$

where Δl_u = the elastic strain of the tendon in its unbonded length l_u

A = the area of steel in the tendon

P = the test load or preload applied to the tieback

E = the modulus of elasticity of the tendon steel

The unbonded length is by definition the tendon length between the structure and the point in the anchor where the tendon begins to be attached. Since the test usually applies load to some portion of the tieback outside the structure (to provide room for jacks), this length has to be considered. The test strain therefore should be:

$$\Delta l_t = \frac{P l_t}{EA}$$

where $l_t = l_u + l_j$ and l_j = the additional length of tendon required by the jacks.

One other characteristic of our test procedures, which should be considered, is the systematic increasing and decreasing of loads during the performance tests. We have concluded that this does not increase the movement of the tiebacks over the movement which would be recorded without the cycling effect. Before concluding that the performance test was the best indicator of the shape and magnitude of the plot of load deformations from the proof test, we ran tests on hundreds of tiebacks. These tests showed a remarkable agreement between the performance and proof tests. Figure 8-5 shows one study from these tests. In this program, a performance test was run on the center tieback in a group of three. Then proof tests were run on the other two tiebacks. The elongation at design load of each proof test was plotted as the ordinate, with the other coordinate being elongation of the related performance test. For this particular project, movement of adjacent

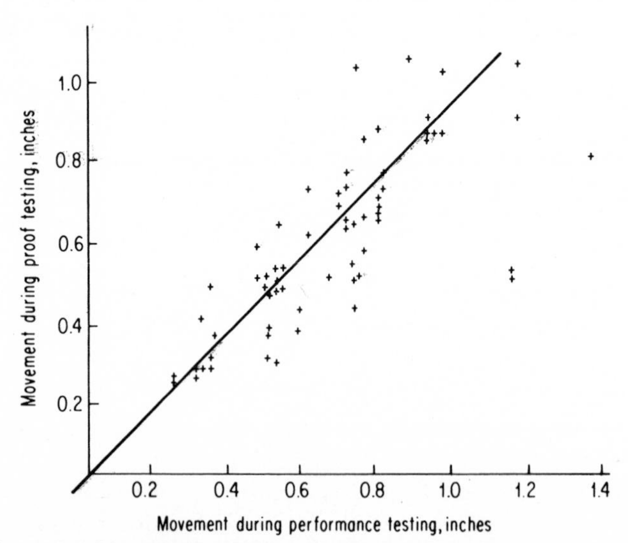

FIG. 8-5 Comparison of movements during performance testing with movement of similar, adjacent tiebacks during proof testing.

tiebacks, where soil and tieback construction were similar, agreed. The difference seldom exceeded $\frac{1}{4}$ inch, and the average variance was practically zero. The tests also showed that agreement was good in spite of large differences in movements due to soil condition, which resulted in a movement range from about 0.4 to 1.1 inches. From these and other similar tests we have concluded that cycling the load does not significantly affect the deformations in our standard testing procedure.

In this part of the chapter I have tried to show that the load/deformation curve resulting from a tieback test is dependent on the way the tieback is designed and also on the soil. As we develop experience with a particular type of tieback, we can learn to anticipate the proper load/deformation pattern. One reason for dwelling on this subject is to stress that, if the acceptance criteria specified are appropriate for only one type of tieback, then all other types are barred. I believe this is done too often and frequently from failure to understand what is occurring.

Loading Errors during Tieback Testing

The results of proof and performance tests are presented as a plot of load versus deformation. Testing errors occur in applying the proper load as a result of jacking errors, seating losses, and other deferred losses, as will be explained later. The deformation measured may be in error as a result of the tendon alignment changing when load is applied, deformation of the structure, movement of the tripod or other fixed point, or failure to recognize a creeping soil. These errors become more significant when tests of greater precision, such as creep tests, are needed.

One of the most common mistakes is to assume that hydraulic jacks and fittings do not leak. Test procedures which attempt to measure the decrease in pressure over a period of time are based on this faulty assumption. So long as jacks leak, are affected by changes in temperature, and have friction, this will be a meaningless procedure. The reasonable alternative is to keep the pressure constant by repumping to compensate for both leakage and tieback elongation and simultaneously measure the tieback deformation. All of the tests I recommend are based on the principle that the test load is maintained by the necessary repumping and deformation is measured.

Jack errors are one source of error in reporting the proper load. It is good practice to calibrate the hydraulic jack with the gauge and pump which will be used with it. Even so, an error of about 3 percent of the jack capacity is not unusual with increasing loads. When the load is cycled, much larger errors may occur as the load is being decreased. This is partly because the jacks are usually calibrated only for increasing load. They would have to be calibrated for load reduction knowing the various loads at which load reduction starts, to be meaningful. The usual way of measuring the test loads is to read the pressure gauge, and even with a calibrated test setup some error should be expected.

Friction in the jack also affects the test results. If the jacks are calibrated, it is reasonable to assume the calibration curve compensates for friction. Thus, when the loads are being increased by jacking, we can assume the measured load is reasonably accurate. I believe the friction has a greater tendency to distort the loads when the load is being maintained without need for much repumping. In this case, jack friction tends to prevent the jacks from responding to slight changes in tieback movement. Friction clearly affects the test results as soon as we decrease the loads. It is common to observe that only a very slight tieback movement occurs when the first portion of the tieback load is reduced. I believe this is frequently because of jack friction. Unfortunately, all jacks have some friction, and therefore all tests are affected by it.

One way of minimizing the effects of jacks on tests is to read the applied loads from a load cell instead of a pressure gauge. The load cell is placed in line with the test jack so that it measures the tieback load directly. In general, load cells can be used to increase the accuracy of reading the test loads, and probably are reliable to about ±1% of the design load.

Seating losses affect the tieback load. These are the losses which occur when the tieback load is permanently transferred to the structure and the jacks are removed. All tieback hardware depends on some seating loss to pull either wedges or nuts into a seat. This loss is common in all post-tensioning practice, but it is minor. It is usually handled by applying to the jacks an overload sufficient to equal both the working load designed in the tieback and the seating loss. Then when the loss occurs, the tiebacks are properly preloaded. It is important that the jacks act to apply loads simultaneously to the structure and the tieback. If this is not done, then the preloading will be affected by structure movement.

The applied preload can also be reduced by various kinds of deferred losses. These can occur from tendon relaxation, anchor creep, temperature changes, and structure movement. Some strands used as tendons seem to slip in the high-bond-stress portion of the anchor. This slippage takes place over a period of days, so that debonding of a portion of the tendon occurs. It seems to occur only in tiebacks with strand tendons. This increases the effective unbonded tendon length and usually results in some load decrease to keep the strain roughly constant. In earlier chapters we discussed anchor creep. Of course, if it occurs, the movement of the anchor toward the structure decreases the tendon strain and results in a proportionate decrease in tieback load. This principle can be used with the creep tests to calculate the anticipated load loss. The preload can then be increased by this amount so that at no time is the tieback load less than the required amount. Of course, temperature can also affect the tieback load. If the anchor and structure do not move, then an increase in temperature decreases the tieback load since the tendon steel increases in length as temperature increases. Usually, temperature changes result in both structure movement and tieback load changes. This is a very complex interaction, but fortunately it is small since most of the tendon is buried in the soil where temperature changes are quite small. Deferred losses frequently result from movement of the structure. It is common practice to determine the preload on the basis of the worst condition, which does not always exist. As a result, the tieback causes the structure to move into the soil until the tieback load and the reaction from the structure are in balance.

Since the purpose of the proof tests is to screen all tiebacks to avoid using one which will not carry the design load, we test to an overload in excess of all the possible errors. The combination of all the losses described above should not exceed 10 percent in a well-designed job, so a higher overload is used for the proof test. Years of experience with a proof-test load of 120 percent of design have shown it is high enough to assure that only tiebacks capable of carrying the load are incorporated in the work.

In the last few pages we have been discussing the causes of error in applying the test loads. These have included errors resulting from the jacks such as calibration error, friction, and alignment. Seating losses, including failure to apply the preload simultaneously to the structure, were considered next. Deferred losses from such causes as creep of the anchor, debonding of the tendon and tendon relaxation, temperature changes, and structure movement were then discussed. These losses are all relatively small, and a reliability coefficient is used to select the overload for testing. Many years of experience have shown that the 20 percent overload will more than compensate for any and all of these sources of errors.

Two errors commonly made will destroy the value of the tests. One is to use a jack with insufficient travel, so that at the end of the test it cannot elongate the tendon. The other is to test strands individually. In the first case, some jacks when fully elongated will show a pressure increase due to their internal construction. I have seen numerous supposedly good tests where the force never was applied to the tieback. Obviously such a test is worse than useless: it is misleading. Testing each strand individually is also poor practice. The purpose of the test is to verify the anchor capacity and unless the full test load is applied to the tieback, we cannot accomplish this. It is very difficult to test the anchor by applying partial tendon loads to separate strands. The load may be ultimately applied to the tieback in this manner, but it is very time-consuming, inefficient, and inexact. It is better in all cases to obtain jacks capable of doing the job and use them to do the job properly.

Measurement Errors during Tieback Testing

A second class of errors in the testing results from reporting the tieback movements inaccurately. These come from seating errors, structure movements, movements of the reference, creep, and failure to consider the jacking length. In general, the movements can be read with a greater degree of precision than the test loads. If reasonable care is used, they should be able to be read to 0.001-inch increments. They are read with either dial or vernier-type gauges. In either case, the accuracy of the movement readings will be more reliable than the load readings. Still, it is important to consider the common causes of errors and minimize them.

One common cause of errors is to overlook the fact that the tendon alignment changes when it is tensioned. This means that the gauge has to be positioned to anticipate this. We put a plate on the end of the tendon so that the gauge will still be in contact if the tendon shifts laterally several inches under load. To overcome this problem, many engineers apply a low load, less than 10 percent of the test load, before beginning to take measurements.

Measuring the jack elongation can be very misleading. Both tieback elongation and movement of the structure occur during testing. Add to this elongation to seat the jacks, and it should be clear that this is not a measurement of tieback deformation. We have had some projects where the tieback load caused the structure movement to be two or three times as great as the tieback elongation. For example, tiebacks which are designed for the load at full-depth excavation may be tested against only the top few feet of a sheeting system. The tieback load at this stage is many times the earth load on the sheeting and tends to pull the sheeting into the earth. The measurements are also meaningless if the reference point is attached to the structure, for the same reason.

The reference point we prefer is a tripod placed far enough from the structure. If it is specially designed so that the gauges can be attached to it, we can then read the movements directly.

Failure to recognize creep is another cause of poor readings. If a tieback is to be used for any reasonable purpose, it cannot creep excessively. If it appears the tieback will be subject to creep, then the readings should include both time and movement. Special care is also required to hold the load constant and read each movement when the load has been restored to its proper value. The test procedure should include a procedure to be followed, such as that recommended in Chapter 4, when creep is encountered in a test. Figure 8-6 shows the pattern of movements to be expected when creep is occurring. It is very

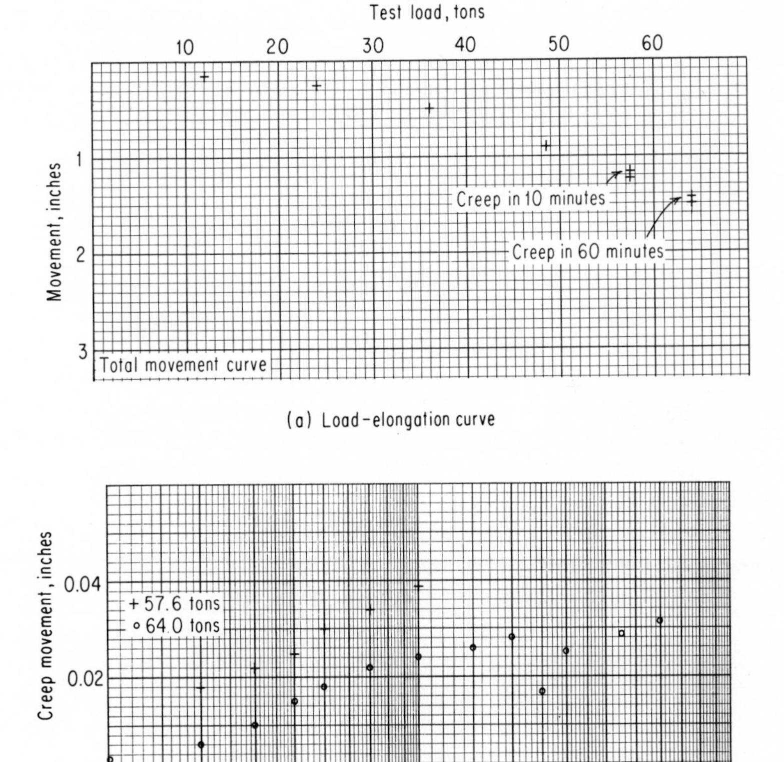

FIG. 8-6 Typical load displacement curve for tieback experiencing creep: (*a*) load/elongation curve; (*b*) elongation/time curve.

obvious from Figure 8-6*a* that the magnitude of movement resulting from anchor creep is very small when related to the total tieback movement. Figure 8-6*b* is included to show one way that the creep movement can be recorded. Creep will be discussed in more detail in the next portion of this chapter.

The length of tendon which is free to strain elastically is usually greater than the unbonded length because some extension beyond the structure is required for the jacks. This length should be recorded at the time the test is made, since it will of course contribute to total tieback movement.

In summary, we can perform very satisfactory tieback tests but we have to recognize and avoid practices which will distort the tests. It is foolish to insist on "perfect" jacks at a jobsite when we cannot even obtain them in our laboratories. We can take certain steps to increase the precision of tests and should know when it is desirable to take them. Finally, I believe the tests are more meaningful if they can be correlated readily with tests on other projects and strongly recommend that the procedures recommended in Chapter 4 be followed.

Creep

Creep is slow movement of the anchor in the soil. It occurs in clays and some silts. At this time, the common approach is to design the tieback so that the rate of creep is low enough that the preload capacity will not be reduced, during its service life, below the design load. To do this, we want to know the rate of creep. We also want to be sure that the rate is decreasing. This is a complex problem which is still being studied. I am sure that in years to come more understanding will lead to better answers. This discussion is based on our current, limited knowledge. What we lack is the confidence that comes with broad experience and long-term success.

Most approaches to designing for creep draw heavily on the theory of consolidation. For a bearing-type anchor, this is a very similar situation. For a shaft-type anchor, we have to consider that maybe the creep will increase with time. In any case, the basis of most designs is that under constant load a clay will continue to consolidate for a long time, but that, by observing the rate of consolidation for a short period, we can extrapolate for the life of the structure. This principle is applied to the creep testing. A constant load is applied to the tieback, and the time-dependent movements are recorded. As discussed earlier, this requires more precise load-monitoring, constant repumping to keep the load constant, and systematic readings of measurements.

The goal of all this is to predict the creep rate and provide for it in the design. The simplest way to do this is to overload the tieback, so that the anchor creep reduces the load over a period of time to the design load. Another way is to return to the tieback several times and restore the load to the design load. A version of this is to run lift-off tests on selected tiebacks to verify that they have not lost any capacity and only restress the tiebacks if selected lift-off tests show creep losses.

For years we have used a simple version of this approach. If the tieback exhibits creep, we hold the test load until the time/movement readings indicate the movement will stabilize with time and also that it will not exceed 0.01 inch in a 5-minute period. This is a simple test which depends on the judgment of the engineer. It has been successful in assuring that the tieback will carry the design load, but it does not provide any means for quantifying the movement to be expected.

The German code requires a more careful test, which should help predict the movement over the life of the anchor. In general, this is the creep procedure recommended in Chapter 4. The problem with this test is that it is extremely difficult to maintain uniform test load, and the measured movements are extremely small. Special pumps, special jacks, and load cells designed to accurately measure load change are required. For example, our constant-load pumps cost twenty times as much as our usual pumps, and we know that they do not maintain constant load with the desired accuracy. This is because they are activated by the hydraulic pressure and cannot adjust for friction in the jack. The load cells must be combined at this stage of the state of the art. When properly done, this test results in a series of load/elongation curves plotted on semilog paper. The movement is expressed as a rate per decade of time. That is, the rate from 1 to 10 minutes is one decade, as is the rate from 10 to 100 minutes. If these are similar, we then extrapolate for the length of the project.

The problem with this procedure when applied to shaft-type anchors is that movement may weaken the soil, not strengthen it as in the consolidation test. This means that over a period of time, as anchor creep continues, it has a tendency to soften the clay. We do not have much factual data on this, but certainly the laboratory tests that have been done so far would lead us to consider it.

Creep is defined as any continuing movement under constant stress. Experience has shown that the rate may decrease with time, remain constant, or accelerate. The rate of creep is influenced by the stress. Below a certain threshold stress, there may be no creep. Above a critical stress, the rate of creep may increase rapidly in comparison to that below the critical stress. Although the causes of creep are complex and only partly understood, full-scale tests under field conditions can indicate whether creep will occur and whether the rate is likely to be constant or increase with time. This information is sufficient for design.

Many tiebacks are in service in soils where the anchor creeps. In some countries codes already specify how to evaluate creep. Many other tiebacks use criteria developed by the engineer or the tieback specialist. It will be important to compare these techniques for evaluating creep and develop clear, rational, effective procedures. I do not think we have done this yet. Instead, we have many procedures, which basically try to estimate the rate of creep to be sure it will not reduce the tieback load below its design load. Some of these procedures are most effective with particular tieback methods. I believe these procedures will be most effective if we can develop a generalized creep-testing procedure which is applicable to any type of tieback, is based on sound geotechnical engineering, and can be readily performed in the field. No such test now exists, but more experience should allow us to evolve one.

Preconstruction Tieback Testing

Tieback tests are occasionally run to assist in the design of the tiebacks for the project. Usually these are made to evaluate the anchor capacity or creep. When the capacity of individual tiebacks is critical to the design, it may be desirable to move onto the site and install and test some test tiebacks. This is particularly true when either the required capacity of the tiebacks exceeds local experience or the required construction method is unusual. In general, the tests can be run as a separate, early testing program or as the first step in the tieback project. They cost more if they are done early because of duplicating mobilization costs.

When the purpose of these tests is to determine the ultimate capacity of the anchor, it is important to use a stronger tendon. This usually means more strands or larger bars. Since yield and ultimate strengths of post-tensioning bars and strands are close and design strength is close to 60 percent of these, it should be clear that the tendon will fail when its load is about (design strength)/0.6. This means that even a test load of two times design load would cause the tendon to fail. Since the test is not really concerned with tendon strength, tendons in test tiebacks should be oversized based on the maximum overload anticipated.

It is also important to select a proper testing jack. The loads and elongations will all be larger than for a performance test, so the jack must be capable of the maximum desired test load. It should not run out of extension before the tieback fails.

In our experience, these tests have been run to evaluate anchor capacity. The best way to run them is to follow the procedure for performance tests but, when the test load for the performance test is reached, to keep increasing the test load in a similar way until the tieback fails. From this failure load, a safe design capacity is selected.

Preconstruction tieback test programs are exceptional. Perhaps they are not performed more often because of the ease of testing production tiebacks, the inherent error in projecting results of a few tests, the frequent inability to duplicate conditions of the production tiebacks, or the high cost. Two reasons we are unable to duplicate conditions of the production tiebacks during testing are our inability to place the test tiebacks in the proper soils and our inability to obtain sufficient tendon capacity. Since tiebacks are frequently used in conjunction with excavation, it may be difficult to install a test tieback in the exact location of the project tieback. Then the test results are subject to more interpretation. Inability to obtain enough tendon capacity is another problem. Many pressure-injected tiebacks are constructed with the largest tendon compatible with the system. This means that these tiebacks cannot be tested even to an overload of two times design load. Since they have a long-shaft anchor, which is frequently presumed to transfer load at a uniform rate per lineal foot, one way the test to failure is obtained is by testing shorter anchors. The rate of load transfer per lineal foot can be obtained, and then the length of anchor selected for the desired factor of safety. Another way is to increase the tendon size, but increase the anchor diameter to account for it. In one case, we test a shorter anchor to failure and in the other we test a larger-diameter one. In neither case are we really testing a tieback with a production-type anchor. Some of the test work that has been done indicates both anchor length and diameter cannot be simply extrapolated, so, after performing the test, we are left with only judgment to interpret the results. Clearly a testing program which is performed on production tiebacks avoids this dilemma.

Tests to Verify Longer-Term Tieback Performance

Load cells, lift-off tests, and various forms of movement readings are frequently used to verify the satisfactory long-term performance of tiebacks. Surveys with levels and transits, extensometers, and slope indicators are used to check on movements of tiebacks in service. Used in various combinations, they can be a very effective control.

One of the most effective combinations is to use load cells to measure tieback load and extensometers to measure movement of the tieback. The load cells are donut-shaped so they can slide over the tendon and under the lock-off hardware. In this position they directly measure the tieback load applied to the structure. It is easy to do this, and many tiebacks have been instrumented in this way. A change in load should mean either the structure or the anchor has moved. We can best interpret the significance if we can also measure the movement. Extensometers allow us to do this with an accuracy approaching a thousandth of an inch. They are quite simple in concept. A frictionless bar or wire is attached to a fixed point. To be useful, it should be oriented so movement is easily read, and this usually means at a right angle to the structure or coaxial with the tendon. We often use a smooth bar in a greased pipe, which extends through the portion of the soil where we expect movement, and then grout the bar into the deeper soil we feel will not move. Using the end of the bar as a fixed point, we read the movement of the structure or tieback. Extensometers are of course not feasible on every project, but in most cases optical surveys are suitable. If optical surveys show the structure has been pulled into the earth, we would expect a reduction in tieback load. Usually the earth loads cause the structure to move some, and optical surveys give a control over acceptable movement. This can then be related to the measured tieback loads.

Another method of checking tieback load is to periodically retest the tieback. Test jacks are reinstalled on the tieback, and loads applied to the tieback by reacting against the structure. At low loads, the jack does not increase the stress in the tieback. As soon as it begins to, an abrupt change in the rate of strain is observed as the stress in the full unbonded length increases. This causes the nuts or wedges to loosen. The load which causes this to happen is recorded as the lift-off load. This is then compared with the load originally locked-off in the tieback. To interpret this, we usually also need movement measurements.

Slope indicators are another tool for measuring movement. The system depends on a track, which is cast into the soil or structure, and a tool for measuring movement of this track. Figure 8-7 shows the results of slope indicator readings on a tied-back sheeting wall. These readings can often be correlated with the changes in tieback design load. In this way, they are a part of the review of tieback performance.

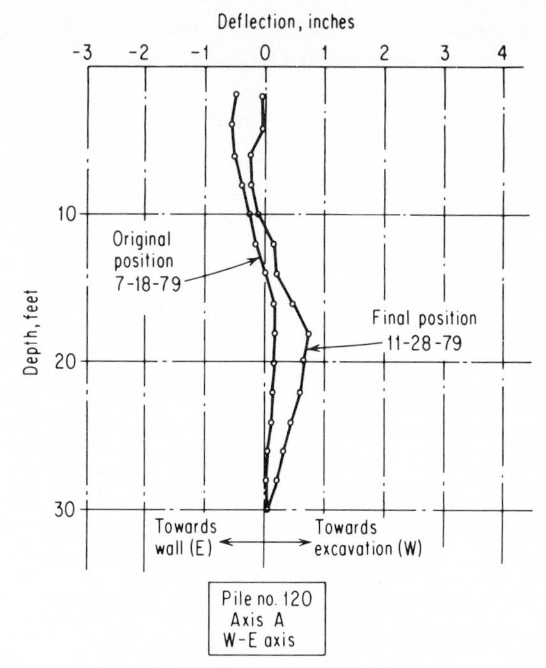

FIG. 8-7 Typical slope indicator report.

The example also shows that they have greater accuracy than the optical surveys, so they can be an effective monitor of total structure performance too.

This review of long-term testing procedures has considered methods of measuring both tieback load and movement. There are certainly other ways of doing it, and this is not intended to be an all-inclusive list. These are techniques frequently used with good success.

Conclusion

Chapter 4 proposed standardized proof and performance tests. These tests verify that every tieback used will carry its design load. In this chapter, we cannot stress too much that the purpose of these tests is to verify the load-carrying capacity of each tieback. A lengthy discussion of creep testing was included because tiebacks which have anchors that creep will lose capacity with time. Methods of testing for creep were discussed, as well as methods for providing for creep in the design. The standardized tests were reported by the use of a load/deformation curve, and a portion of the chapter was devoted to interpreting these. Creep and load deformations were particularly discussed because in both cases invalid criteria may be generalized from particular cases. In this book, careful testing of all production anchors is recommended. Occasionally preconstruction testing is required, and this was briefly discussed. The chapter concluded with a review of some frequently used long-term test procedures.

UPLIFT

A very common use of tiebacks is to resist uplift. When used this way, they are often called tiedowns. Water pressure on the bottom of building slabs or tanks may cause them to float, and tiedowns are used to prevent the flotation. Other structures, such as dams and towers, are subjected to forces which cause uplift, and the tiedowns resist them. In this chapter I will discuss the use of tiedowns for these common applications. There are many other examples of tiebacks used to resist uplift, which will certainly occur to readers, but the principles of their use are similar.

Water Pressure on Concrete Tanks

Large numbers of tiedowns are used to resist uplift pressures on the bottoms of tanks. Figure 9-1 shows a typical tank in a sewage treatment plant. Usually these plants are built in low-lying areas where the groundwater levels are high. The heavy bottom of the tank normally has sewage above it and water pressure acting upward on its bottom. As long as it is full, it has a net downward pressure. Occasionally it must be emptied as a part of the process, and then the water pressure on the bottom exceeds the weight of the tank. The tank can float and many have. To resist this tendency, tiedowns are used.

The underlying principle of this design is that the tiedowns must

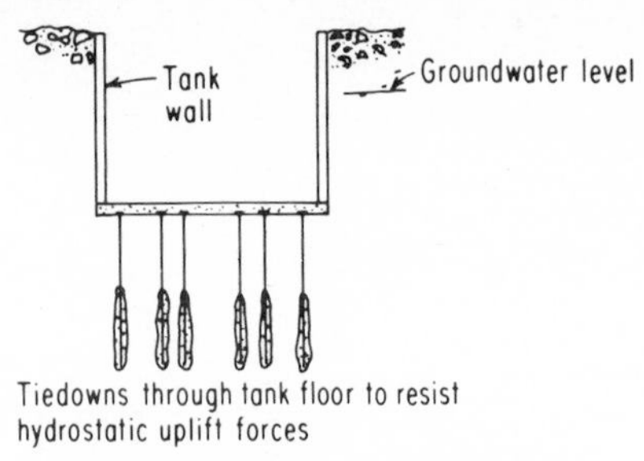

FIG. 9-1 Sewage treatment tank.

attach the tank to a large enough mass of soil. Thick concrete slabs have been used to resist flotation by the weight of concrete, but tiedowns do it much more efficiently by utilizing the weight of some underlying soil. By increasing the length of tiedowns, we can increase the soil mass. We verify that the tiedowns have tied the tank to a large enough mass of soil that it will be stable, even when all the tiedowns act on it. We call this checking mass stability.

The tiedown design must consider both the capacity of individual tiedowns and mass stability. This means that each tiedown should have an anchor in soil or rock which will not pull out and, when all the tiedowns act as a group, they will not raise the soil mass in which they are anchored. For a tank tiedown, it should be clear that each tiedown must be anchored in a mass of soil large enough to resist flotation. The tiedown design determines the size of the mass of soil in which the tiedowns are anchored, so we will consider requirements for mass stability before the design of the individual tiedowns.

The tiedowns are used to hold the tank down by anchoring it to the soil beneath it. Of course, only the buoyant weight of the soil or rock can be utilized, and the tiedowns must be long enough to mobilize a large enough mass of soil. Figure 9-2 shows one way of calculating this. The buoyant force on the tank is calculated. Then a mass of soil large enough to resist this force, with an adequate factor of safety, is selected. The tiebacks' lengths are then calculated so that they will engage this mass of soil. In Figure 9-2, the mass of soil is shown shaded. The buoyant weight of this mass of soil is available to resist the water pressure on the bottom of the tank. In many stratified soils and with close tiedown spacing, it is reasonable to expect a failure on some preexisting surface in the soil. In most shales this would be a reasonable surface also. This surface would usually be at or just below the end of the tiedowns.

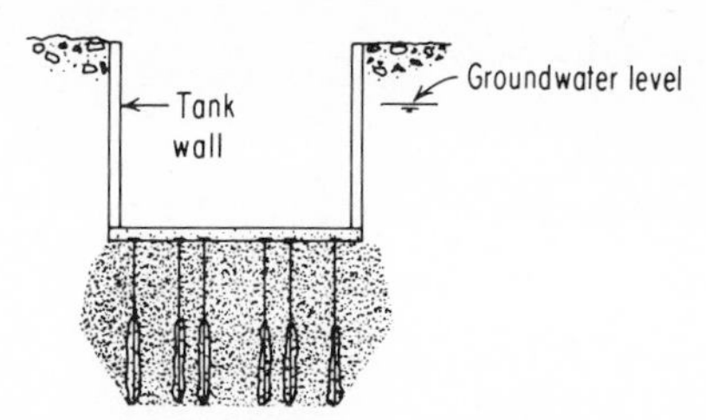

FIG. 9-2 One method of determining mass of soil resisting hydrostatic uplift.

In other cases it is not reasonable to expect the soil to fail on a plane through the bottom of the tiedowns. Calculations are then made on the most reasonable mechanism of failure. For example, if we assume, as in Figure 9-3, that the anchors at pullout will engage the soil around them and tend to lift an inverted cone shape, they may have to be deeper to develop the same mass stability. As can be seen, the weight of this soil mass will be less than the first case. We can also see that, to increase stability, we simply use longer tiebacks, which increase the weight of the soil mass.

Each tiedown must also have an anchor in the soil of sufficient capacity. The design of the slab for the condition of an empty tank establishes the capacity which each tiedown must have. As we discussed earlier, the designer should consider the methods used to make tiedowns on similar projects and follow the methods which have been successful. He should also analyze the soil in which they will be made and select a reasonable design load based on both the anticipated construction method and the soil. This should then be utilized to analyze the slab. I believe that all of the tiedowns used should then be tested during the construction.

Testing, locking off, and corrosion-protecting the heads of tiedowns used for a tank should be carefully considered. The tiedowns are only

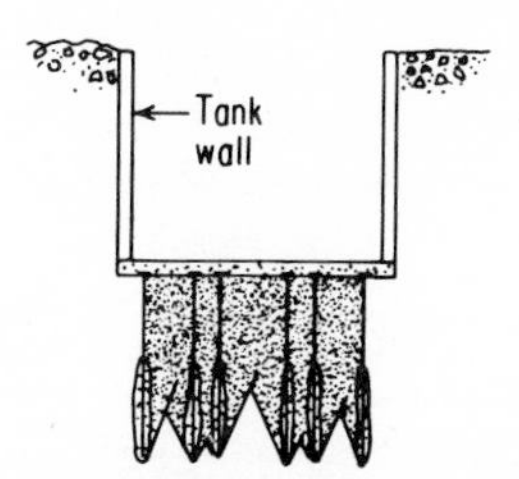

FIG. 9-3 Second method of determining mass of soil resisting hydrostatic uplift.

required to function when the tank is empty. When it is full, it has a net bearing pressure on the soil. If we test and lock off the tiedowns using the bottom slab before the groundwater pressure is acting, subsequent construction will cause the slab to settle and relieve some of the locked-off load in the tiedowns. Allowing the groundwater to rise to its design level outside of the tanks after they are completed but before filling them will create the condition for which the tiedowns were designed. If tank uplift occurs when this is done, we can presume the tiedown loads have been increased beyond the lock-off load. This increase in load is a function of both the tank heave and the unbonded length of the tiedown. By making the tiedown long, and thus increasing its unbonded length, we decrease the magnitude of load change resulting from slab movements. We can also adjust the lock-off load to the calculated load at this time. After locking off the final load in the tiedown, the connection to the slab must be protected. This is usually done by making the connection in a recessed pocket, covering the end of the tendon and its nut or wedges with corrosion-resisting material, and filling the pocket with concrete.

When the tank is filled, its weight is increased, it settles, and the tiedown load is reduced or eliminated. The tiedowns must be selected so that this periodic reversal of stress can occur without reducing the capacity. Most tiedowns for this application are made with their anchors in sand or rock. If their unbonded length is long enough, they can be designed so that the tank slab movements have only minimal effects on the load locked off. For example, with a $1\frac{1}{4}$ rod and ASTM-grade 722 steel,

$$\Delta T = \frac{\Delta l A E}{l} = \Delta l\left(\frac{AE}{l}\right)$$

where A = area = 1.25 in²
E = modulus of elasticity = 30×10^6 lb/in²
ΔT = change in tiedown load from slab settlement
Δl = slab settlement

$$\Delta T = \left(\frac{1.25 \times 30 \times 10^6}{20 \times 12}\right) = 15{,}600 \text{ lb}$$

change in tiedown load from slab settlement of 1 inch where the unbonded length is 20 feet.

$$\Delta T = 7800 \text{ lb}$$

change in tiedown load from slab settlement of 1 inch where the unbonded length is 40 feet. If we can analyze possible settlement and heave of the tank bottom, we can properly adjust the tiedown loads.

Building Slabs

In most cases, slab foundations used for buildings have a positive bearing pressure, resulting from the weight of the building. The slab is usually designed to span between columns. If the column load exceeds the hydrostatic pressure on the particular column bay, no tiedowns are required. They are used where the structure is too light to overcome this hydrostatic uplift. They may also be used to reduce the spans on the slab and thus its cost, even where the column loads are heavy enough. Figure 9-4 shows a typical application of tiedowns to a building slab.

Individual tiedown capacities are usually in the range of 50 tons. Spacing is related to column spacing, usually 5 or 10 feet on centers. Unlike the tanks discussed earlier, the load is usually not reversible. However, the settlement of the slab during construction will definitely affect the load locked off in the tiedown.

The tiedowns should be designed so that they are anchored in a mass of soil which would not be displaced by the group action of the tiebacks. The design principles are similar to those for the tank foundations. Leakage through the slab around the tiedown is a more important consideration for this use. Coordination with other phases of the construction may also be more difficult.

It is almost impossible to prevent some leakage around a tiedown which penetrates a slab on which large hydrostatic pressure acts. For some slabs, this may not be a problem. The slab may be used as the floor of a parking garage, for instance, in which case a little dampness may be acceptable. If dampness is not satisfactory, one of two procedures is usually followed. The tiedown may be designed to be installed, tested, and locked off before the slab is poured and have the slab poured around it. Figure 9-5 shows some tiedowns designed to be used in this way. When the work mat was poured, thickened sections with reinforcing were provided at each tiedown location. Then the tiedowns

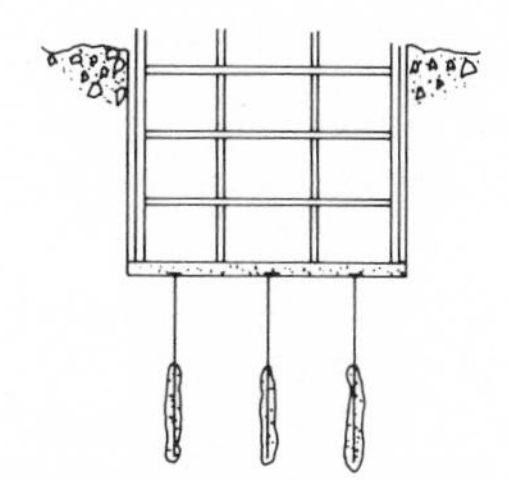

FIG. 9-4 Tiedowns, used with a building slab, replace a thick reinforced mat with a thin slab and permanent tiedowns.

FIG. 9-5 Tiedowns designed not to penetrate a building slab.

were installed off of the work mat. They were tested and locked off against the work mat. A portion of the tendon extended above the work mat into the slab. A plate and nut were placed so that they would be poured into the mat. In this way, no hole through the slab is required, and the possibility of leakage from this cause is eliminated.

The other way of assuring a dry lower slab is shown in Figure 9-6. The slab is a structural slab through which the tiedowns were installed. They were then tested against the slab and locked off. Then a foot or so of gravel was placed on this slab, together with drains and sumps, so any water which leaks into this space is drained away. The floor slab, of nominal thickness, was then poured on the gravel. In this way, leakage through the structural slab can be tolerated, without detracting from the use of the building.

The first two sections of this chapter have considered tiedowns used to resist hydrostatic uplift forces on tanks and building slabs. In general, the tiedowns must be anchored in a mass of soil which is large enough that its buoyant weight exceeds the hydrostatic uplift. Each tieback must be capable of developing its anchor in this soil mass. The next two sections will consider tiedowns used to resist uplift forces on dams and towers.

FIG. 9-6 Tiedown installed through a building slab.

Tiedowns to Strengthen Dams

A gravity dam is shown in Figure 9-7, with a tiedown. The tiedown strengthens the dam by increasing its resistance to overturning and sliding. The water pressure creates a moment M_w which acts on the dam and its foundation. Of course the concrete of the structure is analyzed, but to strengthen the dam we usually apply a moment, which acts opposite to the overturning moment caused by the water,

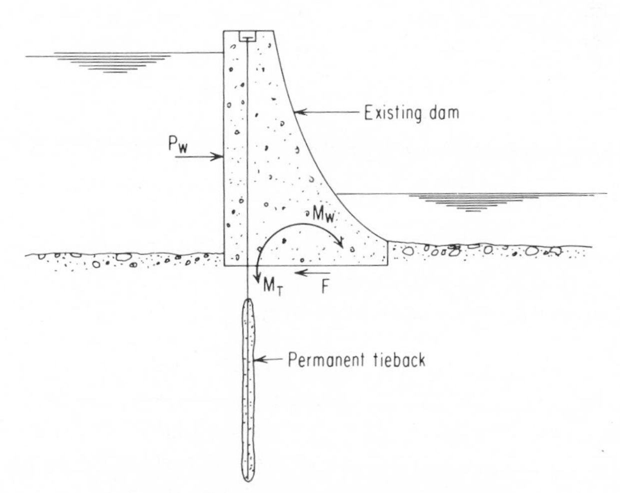

FIG. 9-7 Tiedowns used to strengthen a dam.

and increase the resistance to sliding by increasing the normal force. The dam design is first analyzed to determine the desirable induced moment from the tieback. Since the moment caused by the water pressure is usually resisted by bearing, it is important that it not be large enough to cause tension beneath the upstream edge of the dam. Since the moment also causes a maximum bearing, which is about twice the average bearing, near the toe of the dam, the dam stability may also be limited by the available rock bearing pressure. If the moment caused by the tiebacks is large enough, we can assume that tension does not develop at the upstream edge of the dam and lower the bearing at the downstream edge.

The tieback should be near the upstream edge of the dam. It provides a moment resistance which can be calculated when we know its distance from the centroid of the foundation. We can calculate the required tieback capacity when we know the moment and lever arm. The moment requirement is usually developed from an analysis of the dam, the forces acting on it, and the required factor of safety. Calculation of it is beyond the scope of this book, but the principles should be clear.

The resistance of the dam to sliding is often the friction on the base, which is the product of the normal force and the coefficient of friction between the dam and rock. If the normal force is increased, which the tiedowns do, the friction on the base is increased.

The friction F is calculated:

$$F = N \tan \phi$$

where F = frictional resistance to sliding between the dam and its foundation

N = normal force on the foundation

$\tan \phi$ = coefficient of friction between the dam and its foundation

This example is greatly simplified to illustrate the point, but it should be clear that increasing N, by tiedown forces, increases frictional resistance.

With tiedowns used to strengthen a dam, the usual solution is a row of tiebacks near the upstream face. Figure 9-7 illustrates this. The tieback hole is first drilled through the dam, then into the soil or rock beneath. In most cases, this type of dam is built on rock, so the drilling equipment is suitable both for rock and concrete.

Tiedowns anchored in rock are easy to make. The hole can be drilled with percussion tools. For shallow holes, rock drills on air tracks are used. For deeper holes, roller cone bits or down-the-hole hammers are usually used. The rock-grout bond is high, so large capacities can easily be made. Dams are massive structures which can accept large forces from the tiedowns. The ends of the tiedowns can also be easily protected by recessing them in the dam. The tiedowns provide an ideal way to increase the stability of a dam.

The rock, in which the tiedown is anchored, must hold it. The bond between the anchor and the grout must not be exceeded. We can test individual anchors to verify this. We also have to analyze the anchors to be sure they will not lift the entire rock mass. This condition is shown in Figure 9-8. Even if individual tiedowns are successfully tested, the system could fail. The failure would be a mass-stability problem caused by the tiedowns moving the mass of soil in which they are anchored.

The tiedown ultimate capacity cannot exceed the weight of the rock which would break free when the tiedown is loaded. Tests of individual tiedowns can be made to verify that the anchor is capable of transferring the desired tiedown load to the rock. Mass stability, on the other hand, must be confirmed by calculations.

Engineers have used various methods to calculate the shape and weight of this mass. The shape of the rock mass is assumed by some to be conical, with the tip of the cone at the bottom of the tieback. Where adjacent cones overlap, vertical cylinders which butt one another are substituted. Due consideration is given to buoyancy, and the weight of the rock in the cones or cylinders is calculated. The tiedowns are analyzed for group action, and the lengths in rock calculated so that the weight of rock in the cones and cylinders exceeds the ultimate tieback

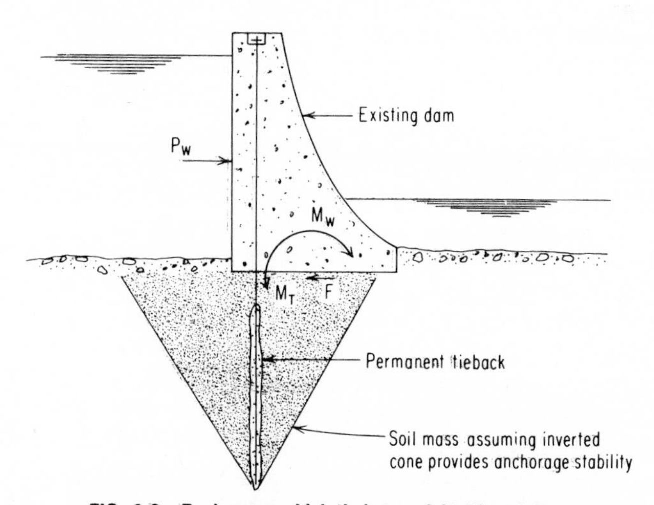

FIG. 9-8 Rock mass which tiedown might lift at failure.

capacities. In this way, the dam is tied to the rock beneath it, instead of only bearing on it. Some of the rock deadweight is added to the dam's deadweight to increase its stability. Tests have shown that the tiedown pullout capacity calculated by the cone method is grossly understated. Rock, after all, has shearing and even tensile strength. As a result, the mass of rock which resists lifting is larger than the cone calculated in this way. The shape is largely fixed by natural joints in the rock and the size of the resulting blocks. If the engineer has enough confidence in his rock-mechanic's ability and the available data, he can analyze the block this way. To assure the integrity of the tiedowns, he must determine that the block of rock which they tie to the dam is big enough to accomplish the desired purpose. The length of tiedown in rock is increased to increase the size of the block. The length in rock must be long enough both to transmit the tiedown load to the rock in bond and also to tie to the dam a large enough block of rock to resist overturning. Usually the second condition controls. For widely spaced tiedowns, the increase in the rock weight is roughly related to the third power of the length increase. Since much of the drilling is through the concrete of the dam, it is economical to use large-capacity tiedowns. A mass concrete dam is ideally suited to accept these large, concentrated tiedown loads without overstressing. As a result, tiedowns with large capacities have been widely used to increase the strength of existing dams.

Tower Uplift

The last example of tiedown applications is to resist uplift of tower legs. Years after it was completed, a television tower was reanalyzed for uplift. The triangular tower had three legs, each of which rested on a deep concrete footing. The footing bearing was on rock, so the capacity in bearing was more than adequate. The footing had a pier which extended to the ground surface. The uplift was provided by the concrete weight of the footing and pier, plus the weight of the soil fill on top of the footing. When it was concluded that the tower, under severe wind loading, would need more resistance to uplift, tiedowns were installed to overcome the calculated deficiency.

Two tiedowns at each leg were sufficient. The tower engineers calculated the uplift which had to be resisted and found that the best place to apply the resistance was to the pier just below where the steel column had its bearing. A collar for the pier was designed, and the tiedowns installed from the existing ground surface. A hole was drilled through the fill and through the footing, and a casing was set. Then a hole was drilled far enough into the rock to make the tieback. The length, as in earlier cases, was analyzed to anchor the tower leg to a mass of rock with at least enough weight to assure a stable mass.

Summary

In this chapter we have discussed the application of tiedowns to several types of structures to prevent uplift. We have used tiebacks for all of these applications and many more. They tie a structure to a mass of soil or rock to prevent it from rising, and it is important in the design phase to make the tiedowns long enough that this mass will be stable when the tiedowns are loaded.

10

TIED-BACK WALLS

Tiebacks, which are anchored in the soil behind a wall, are frequently used to resist all or most of the earth pressure acting on that wall. This is the most obvious and most often used application of tiebacks. Engineers generally agree that failure of this type of wall results in movements of a wedge of soil, similar to the wedge shown in Figure 10-1.

In the next chapter I will discuss the special walls built to stabilize landslides. The distinctions between these two types of walls—normal earth-retaining walls and landslide-correction walls—are significant enough that a chapter has been devoted to each. Having made the distinction, I want it to be clear that throughout the rest of this chapter, when I discuss walls, I am referring only to the first kind, not to landslide-correction walls.

Tiebacks have been used with various methods of wall construction. Some walls are built before excavation starts, some as excavation proceeds, and some after excavation is completed. Some are designed only for short-term use, as, for example, excavation bracing. Most are designed for the pressures needed to assure stability, but others are designed for larger earth pressures, usually in the belief this results in less movement of the wall and in the earth wedge behind it. The method of wall construction, the slope of the contact between it and the earth it retains, and even whether it is temporary or permanent may affect the design earth pressures. The wall and tiebacks must act together to

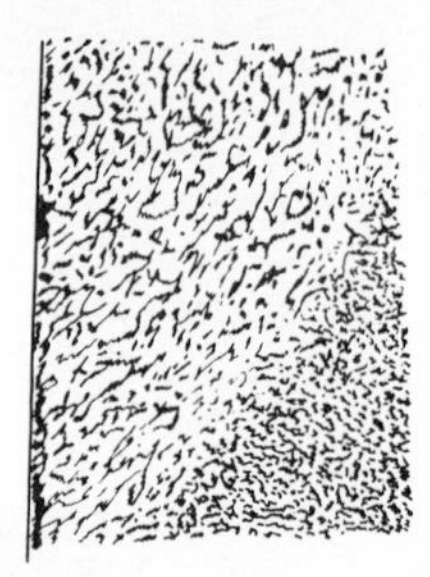

FIG. 10-1 Wedge of soil affected by failure of model braced excavation.

restrain the earth and resist the pressure it applies on them. The tieback design must consider not only capacity but also length so that the anchor is dependable. Finally, tests are available which verify the tiebacks will carry the design loads and also that the design loads are larger than the actual loads. Each of these factors—selection of the type of wall, earth pressures, wall design, tieback design, tests to verify capacity of installed tiebacks, and tests to verify the adequacy of calculated tieback capacity—must be considered when a tied-back wall is used.

The Wall Type Affects the Earth Pressure

All of our experience has taught us that the method of constructing the wall can affect the pressures which act on it. Before proceeding to consider some representative types of tied-back walls, we must recognize that we can affect the pressures acting on a particular wall by our selection of the way the wall is constructed.

Pressure increases with depth against many walls, but not against most tied-back walls. Rankine developed the familiar analytical procedures for walls similar to Figure 10-2, which related the horizontal

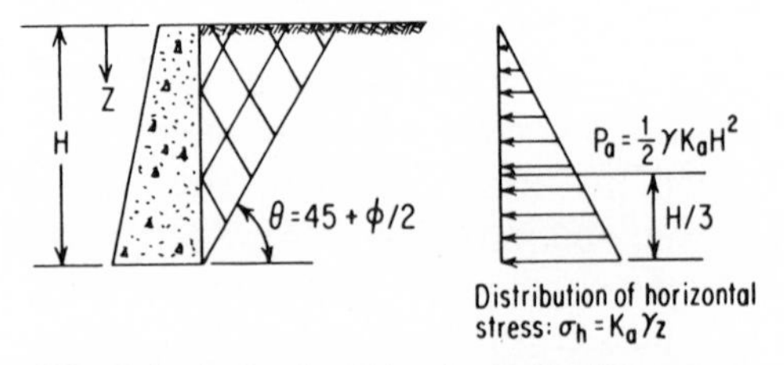

FIG. 10-2 Active thrust for simple Rankine case.

pressure at any point to the vertical pressure: since the vertical pressure increases with depth, the horizontal pressure increases with depth. The analytical conception is a triangular pressure distribution. Of course this is a simplified discussion, but the important point to understand is that the vertical pressure in the soil was one principal stress, and the horizontal pressure was another, although it was a fraction of the vertical. Most walls are designed using this basic theory, and have performed well.

When this theory was used for the design of excavation bracing walls, many failed. Various early investigators found that the top braces failed because their load was higher than calculated. K. Terzaghi explained that the difference in pressure was a direct result of the way in which the wall was built and recommended a redistribution of pressure for design. An examination of Figure 10-3 will show that this "apparent-earth-pressure" distribution had the shape of a trapezoid and a greater total force than the triangular active-earth-pressure design. The figure also shows a later recommendation of uniform pressure against the wall for a sandy soil. Clearly, each applies higher pressures to the top of the wall than the Rankine procedure. Terzaghi recommended this trapezoidal design pressure diagram for walls which are built directly against the excavated earth and explained why the redistribution of the earth pressure into this "apparent-earth-pressure" diagram resulted from installation of bracing as the excavation proceeded. Over the years, most excavation bracing walls have been designed following the Terzaghi procedure, and have performed successfully.

These procedures aim to assure a stable wall. The wall will be stable if it is strong enough to resist the active earth pressure, but it must move slightly. This is basic soil mechanics theory, which states that the horizontal pressure in the soil "at rest" is more than the active and less than the passive. Most walls must move toward the excavation so

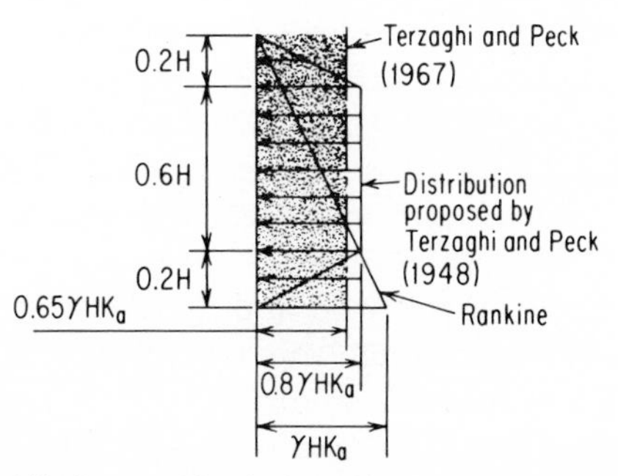

FIG. 10-3 Pressure distribution used for design of bracing system. (Terzaghi and Peck, 1948)

that the "at-rest" pressure can be reduced to the active. A consequence of this understanding is that movements of both the wall and the wedge of soil behind it can be expected if we design for the active case. If we design the wall for higher pressures, we may reduce this movement.

All of the preceding discussion presumes the wall is vertical, but, if the wall is sloped, we know the pressure will be reduced. Even a slight slope results in a large reduction of earth pressures. It is not unusual to batter a wall 10° from the vertical and find the earth pressure reduced by a third.

Water pressure against the wall may also be affected by the wall construction. For example, no water pressure can act on soldier-beam and wood-lagging sheeting, since it leaks so readily. In many soils with a high, or perched, water table, but low permeability, this may be a controlling consideration. An alternate, watertight wall would have to be designed not only to resist the earth pressure, but the water pressure as well. Clearly the type of wall may affect the total pressure which acts on it.

Tied-back walls can be either temporary or permanent, and many engineers design a permanent wall for higher earth pressures than a temporary wall at the same site.

In the preceding general discussion, I have tried to show how the earth pressure can be affected by the wall construction. This interdependence must always be considered when choosing the wall and evaluating the pressures which act on it. It must also be kept in mind in the following separate discussions on various ways of constructing the walls and establishing earth pressures.

Types of Walls

The unifying characteristic of all the walls discussed in this chapter is that they use tiebacks. These tiebacks apply a horizontal force on the wall to resist the horizontal earth pressures acting on the walls and limit the wall deflections and associated earth movements to acceptable levels. The tieback anchors are made in the soil behind the wall, and preloading the tieback applies a force to both the wall and the soil around the anchor, so that the tieback provides a precompression in the wedge of soil between the wall and the end of the tieback.

Figure 10-4 is a section through this type of tied-back wall. Tiebacks have frequently been used with soldier-pile walls, walls of contiguous drilled piers, steel-sheet piling, and concrete diaphragm walls. Most deep underpinning in recent years has also been tied back. One type of

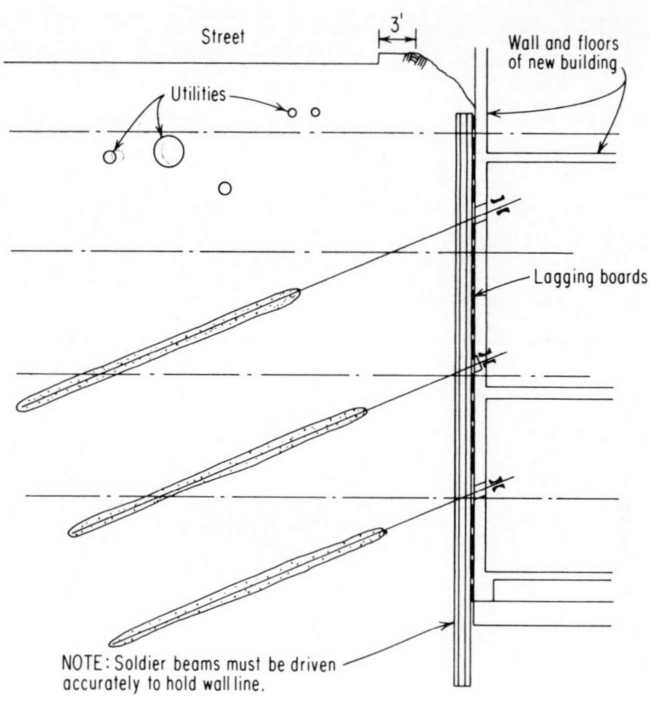

FIG. 10-4 Typical tied-back wall.

permanent wall, which we call a "Georgia wall," combines a concrete facing with tied-back soldier-beam sheeting. Permanent tiebacks have also been used with large cylinder piles or sheet piling for drydock construction and waterfront structures. In many instances they have been used to provide added horizontal forces on walls which were failing.

In the next few pages, I will consider some common types of walls. In each case, the tiebacks apply a force to the wall. The wall, in this discussion, refers to the construction against which the earth presses. I have made this distinction between the wall and tiebacks at this time so that we can briefly consider some of the many ways in which the wall can be constructed. In all cases it must be structurally designed so that it spans between tiebacks. Later in the chapter I will discuss the tiebacks further.

Tiebacks have been combined most often with soldier-pile walls for excavation bracing. The wall in Figure 10-5 shows some common procedures used with this type of wall. The vertical soldier piles are driven bearing pile sections. These piles are the vertical reinforcing of the

FIG. 10-5 Driven soldier-pile wall.

wall. Between piles, the earth is restrained by the boards. The soldier pile is analyzed, and a force calculated which must act horizontally to assure stability. The tieback and wale are then designed to provide this force. The figure clearly shows the tiebacks, wales, boards, and the face of the soldier pile.

If a projecting wale cannot be tolerated, a soldier-pile wall can be built as shown in Figure 10-6. The soldier piles are pairs of beams or channels installed by drilling a hole, placing the steel in the hole, and filling the hole with concrete. Then the tiebacks can be drilled through the center of the soldier piles in the space between the beams or channels.

Another type of wall is the drilled-pier wall, shown in Figure 10-7. In this case, the wall is constructed by drilling large-diameter holes in the ground, placing the reinforcing, and pouring the holes full of concrete. The wall is built sequentially, so that when completed it consists of a row of contiguous drilled piers. The drilled piers are reinforced so that they can span vertically between rows of tiebacks. This type of wall can be used both as a temporary or permanent wall. When used as a

FIG. 10-6 Soldier piles installed by drilling.

FIG. 10-7 Drilled-pier wall.

FIG. 10-8 Tiebacks with sheet-pile wall.

permanent wall, it usually requires a facing and frequently has modified details to cause the piers to overlap.

Sheet piling is used where the driving is easy or groundwater must be controlled. A wale, similar to Figure 10-8, is usually required with sheet piling. When the groundwater level is maintained above subgrade, there will of course be water pressure on the wall, for which it must be designed. Sheet piling or any impervious wall may result in groundwater flowing under the wall and upward into the base of the cut. This flow may cause sand to boil, passive pressures to be reduced by seepage, and other conditions which the design must consider.

Tiebacks have also been used to support concrete diaphragm, or slurry, walls. In this case, the wall is internally reinforced to span vertically and horizontally between tiebacks. In general, this type of wall is constructed in abutting panels, and each panel is individually designed for a pattern of tiebacks required to provide the horizontal force. The wall in Figure 10-9 has been tied back in this way.

All of the examples cited have been temporary excavation bracing walls. In general, these structures have a short life. In some cases, only

FIG. 10-9 Tiebacks supporting concrete diaphragm wall.

months of service are expected. When used to support permanent walls, some changes in the tieback are necessary, as well as the wall. Corrosion protection, the finish of the wall, and the leaking of groundwater through the wall must be considered.

We call the wall shown in Figure 10-10 a "Georgia wall." A concrete facing is combined with a soldier-pile-type wall. The concrete is designed to span horizontally between soldier piles. The vertical bending is carried by the soldier pile, which is restrained by a horizontal force from the tieback. The finished wall looks like other concrete walls, but it has no footing and is totally dependent on the tiebacks to provide horizontal resistance to the earth pressure.

All of the walls discussed so far seem to meet the Terzaghi and Peck criteria for use of the "apparent earth pressure." Some walls are built in the soil prior to the start of excavation, such as steel-sheet piling, contiguous piers, and cylinder piles. Others, such as soldier-pile walls, have the boards installed as excavation proceeds. In all cases, the tiebacks are installed from top to bottom as excavation proceeds. This is the condition which caused the pressure redistribution against the wall, explained by Terzaghi and Peck.

FIG. 10-10 Georgia wall under construction.

One special tieback application is to existing walls which are failing. In this case, it is very difficult to calculate the tieback force which will restore stability. The existing wall may also require buttresses or wales since it is usually not strong enough to span between the tiebacks. Figure 10-11 shows tiebacks used to support a wall which had been failing. In this case, we preserved an old wall, but only by adding a wale to distribute the tieback forces horizontally.

Even a casual reading of the last few pages will show that tiebacks have been successfully used with many types of wall construction. There simply is not enough space here to catalog examples of every type of wall construction which has been tied back. These are examples picked to illustrate the widespread use.

Earth Pressure

After selecting the appropriate type of wall, the next step in the design of a tied-back wall is to select the earth pressure. Since the tiebacks are installed as excavation proceeds, the earth pressure on most tied-back walls should be distributed according to the "apparent-earth-pressure"

FIG. 10-11 Tiebacks reinforce a failing wall.

diagram of Terzaghi and Peck or a reasonable modification of it. As discussed earlier, walls that are sloped may have lower pressures, and design for higher pressures may be considered to reduce wall movements. If surcharges or water exert pressure on the wall, these must be added to the earth pressure.

We use the earth-pressure envelope of Figure 10-12 for the design of most tied-back walls. It is very similar to the 1948 recommendation of

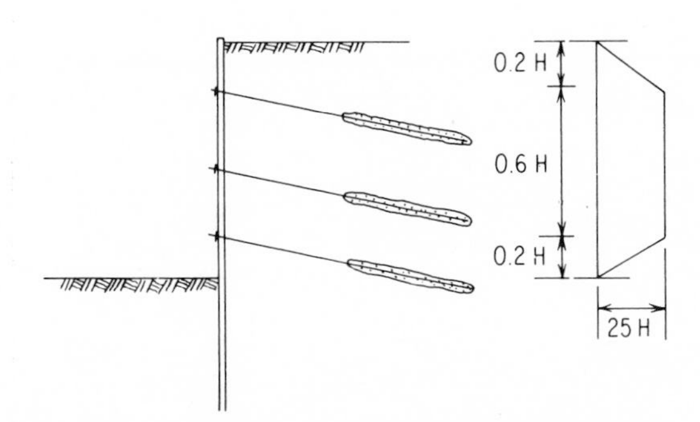

FIG. 10-12 Schnabel "apparent-earth-pressure" envelope.

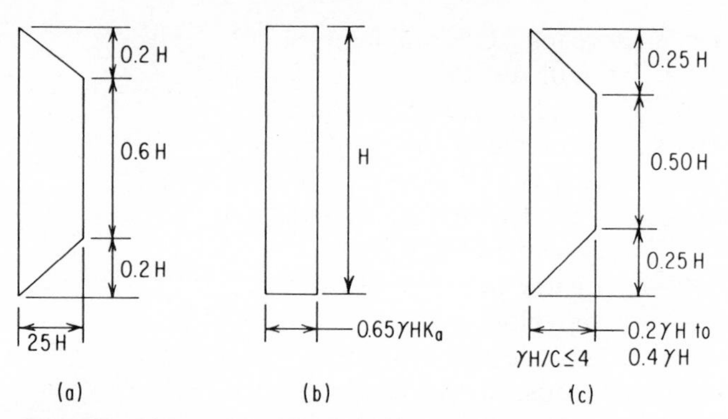

FIG. 10-13 Comparison of several "apparent-earth-pressure" envelopes: (*a*) Schnabel; (*b*) Terzaghi and Peck sand; (*c*) Terzaghi and Peck clay.

Terzaghi and Peck. One difference is that we use it for sand, clay, or mixed soils and don't distinguish between them. Another is that pressure is expressed only as a function of the depth of excavation. For most cases, these do not result in much difference from the Terzaghi and Peck recommendations. To illustrate this, we compare our envelope with the latest recommendations for sand and clay in Figure 10-13. For reasonable soil values, there is good agreement on the maximum unit pressure p and total pressure P against the wall.

Schnabel envelope:

$$p = 25H \quad P = 20H^2$$

Terzaghi and Peck sand envelope: assume $\gamma = 125$ pcf $\phi = 35°$ $k_a = 0.27$

$$p = 0.65\ \gamma\ Hk_a = 0.65 \times 125H \times 0.27 = 21.9H$$

$$P = pH = 21.9H^2$$

Terzaghi and Peck clay envelope: assume $\gamma = 125$ pcf

$$p = 0.2\ \gamma\ H = 25H$$

$$P = p \times 0.75H = 18.75H^2$$

Most of the walls we have designed have been soldier-pile walls. Where soil is good enough that we can build a soldier-pile wall, I believe we can use the earth-pressure envelope of Figure 10-12. We would not use it blindly in softer soils. This is an empirical earth-pressure distribution, based on measuring actual jobs, and we use it to design similar projects in similar soils. Clearly, if data exists on which to refine this approach, it should be used, but I believe in most cases a similar envelope will result.

Two good reasons for expecting lower earth pressures than those in Figure 10-13 are preloading the tiebacks and sloping the wall. When tiebacks are used, it is customary to preload each tieback to a predetermined load. This of course applies the preload to the wall. In this way, each tieback can be accurately stressed to the load which we calculate it must carry to assure wall stability. The total force calculated using these envelopes is about 30 percent greater than the active earth pressure, because the envelopes were originally developed by measuring brace loads. Measured brace loads varied from the average measured load by as much as 30 percent, and this seemed to be the result of workmanship and other factors over which the designers had no control. To be sure the most heavily loaded brace did not fail; the envelope was increased to provide for this random variation of individual brace loads above the average. The total calculated active pressure (i.e., the area of the envelope) is 30 to 40 percent greater than the total measured brace loads. Tiebacks are preloaded, and when they are used, it is probable that this overload to account for randomness is superfluous. If so, an envelope with smaller pressures is reasonable when tiebacks are used instead of braces. Little advantage has been taken of this possible saving.

We have always reduced the earth pressure when we sloped our sheeting. Clearly, if a bank will stand unsupported on some slope, as we approach this slope from the vertical, we expect the earth pressures to decline until they reach zero.

Higher earth pressures are often used in the design to reduce wall movements. A wall will be stable if it is strong enough that it can support the earth pressures which act on it, which are generally considered to be the "active" pressures if necessary movement occurs. Theory explains that a higher "at-rest" pressure existed in the soil on the plane of the wall prior to its construction but, as construction proceeded, small movements toward the excavation reduced these pressures to the "active." These movements are generally in the range of 0.001 to 0.0025 times the depth of excavation and are acceptable for most walls. Naturally, smaller movements will occur in the soil behind the wall. When movements of this magnitude are not acceptable, smaller movements are sought by building stronger walls. One approach is to increase the design earth pressure to "at rest." This results in a stronger wall, with more tieback force on it. Clearly, it should be stable, even with the uncertainty surrounding the determination of the "at-rest" pressure. Observation of many walls has shown that much of the wall movement occurs below the excavation level, prior to installing the braces. Because this happens, total wall movements may not be prevented by increasing tieback capacity.

However, using tiebacks instead of braces may be a more effective way of reducing movements behind a wall than is generally appreciated. The tieback preload is fixed by the loads expected at maximum depth of excavation, but it is applied when the excavation is shallow. It is far in excess of what is needed for stability while the job is being excavated. As a consequence, the wall is stronger than needed for every stage of excavation except the final. The tiebacks also provide horizontal reinforcing in the soil behind the wall and usually below the excavation level, so that horizontal movements in this area are restricted.

When we have determined the earth pressure which will act on the wall, the next step is usually to design the wall.

Wall Design

The unique feature of these tied-back walls is that the tiebacks act to resist all or most of the earth pressure. The walls are designed to support the earth pressure between tiebacks. The wall must also be designed for the concentrated load at each tieback location and any downward force the tiebacks apply to the wall. Design of the wall and tiebacks is interrelated, and a good design will consider this, but for purposes of discussion they are treated separately in this and the next section.

Most tied-back walls are designed to span between rows of tiebacks. The tiebacks are usually installed in long rows at about the same elevation, as is evident from the pictures of walls earlier in this chapter. The tieback capacity must be reasonable for the technique chosen and for the soils at the site. Typically, the wall is analyzed for the required section modulus to support the earth pressure between rows of tiebacks, and sized to provide the required strength.

Clearly, the walls are usually continuous members of constant strength, supported by one or more rows of tiebacks and penetration of the wall below subgrade. We would expect them to be analyzed by applying the earth pressure to a beam over multiple reactions, but this is seldom done. The usual design convention followed is to consider each span between tiebacks (except the top) as a simple span.

Most walls penetrate below subgrade, but no pressure acts on the wall below subgrade when using the envelopes of Figure 10-13. For engineers familiar with bulkhead design, this may seem unreasonable, but there is ample evidence to convince me there is little or no pressure on the walls below subgrade. Certainly, if there is any, it acts on the excavation side and is in the first few feet below the subgrade. We do not apply any pressures below subgrade in designing walls. From this paragraph and the preceding, one might conclude the wall design is inexact, and I would agree. The problem is caused by the fact that the wall deflects between tiebacks and the lower part of the wall yields inwardly as excavation proceeds and causes the soil to arch. If we design the wall and neglect this arching, we will overdesign the wall. The convention and earth pressure are developed together, and still give the most reasonable way of designing the wall. These procedures have been employed to successfully design most tied-back walls.

Among techniques which are used to reduce wall deflections are designing for higher earth pressures, increasing wall stiffness, or reducing tieback spacing. Clearly, all of these affect the wall design. In the first case, the wall must be stronger than required for stability. A stiffer wall may also reduce movements and is frequently the reason for choosing a diaphragm wall over another type of wall. Of course this choice also involves increased costs, so it is usually justified on the basis of superior performance. My personal opinion is that this is one of the most frequent errors in foundation design and is usually made without considering all the alternative techniques. Another way of achieving wall stiffness, if that is the desirable approach, is to reduce tieback spacing. All of these techniques are used in an effort to reduce the movements of tied-back walls, but little or no clear evidence is available to evaluate the cost-benefit ratio.

The tiebacks usually apply a downward force on the wall. Many walls have failed as a result of not considering this condition. Clearly it is important, and should not be ignored when designing the wall. The dangerous force is the component of the tieback in the plane of the wall, and it can be reduced by making the tieback more nearly perpendicular to the wall. In most cases, this means making the tiebacks flatter. With sloped sheeting, the tieback can often be perpendicular to the wall and still be installed at a suitable angle. Of course, since the earth pressures and consequently the tieback preloads are reduced with sloped sheeting, the downward-acting component will always be reduced. Flatter tiebacks reduce the force down on the wall, and this is one method of minimizing the problem. The two common methods of resisting the applied force are for the wall to develop sufficient shear against the excavated earth or to penetrate below subgrade and develop sufficient capacity in bearing or friction. Shear between the wall and earth is the method usually relied on. The normal force on this surface, as well as the shear force, are related to the tieback load. If the friction angle between the wall and soil can be established, the angle between the tieback and the wall can be fixed so that the friction exceeds the shear. When this cannot be done, the wall should be designed to develop resisting capacity below subgrade and carry this component of tieback capacity to the proper depth. This is a particular problem on projects where the wall supports earth above a rock face, since the wall usually stops on the rock above subgrade. Failure of the rock face, induced by the downward load from the wall, can, and has, caused walls to fail. Consideration should be given to carrying some elements of the wall through the rock to below subgrade. For soldier-pile walls, this is often done by drilling holes in the rock large enough to receive the soldier pile. Other, cheaper patented methods are also available. This is not a summary of all available techniques. My intent is only to show that tiebacks apply a force down on the wall, which must be designed for.

In general, as the wall is designed, the tieback loads required to insure its stability are calculated. The wall is designed to span between tiebacks, and the tiebacks must be designed to resist the calculated reactions. Of course, the tiebacks apply a force which acts in nearly the opposite direction to the earth pressure, so they are very efficient for this application.

Design of Individual Tiebacks

The tieback load is computed by analyzing the soil and the wall. The tieback must then be designed to carry the computed load. Its anchor must be made deep enough in the soil that it is beyond the area which would be disturbed by wall movements. The tieback must also be long enough that it will not pull out and the mass of soil it ties together is stable. In earlier chapters, tieback design, construction, and testing were discussed, so only those design considerations peculiar to this type of wall will be addressed in this chapter.

To calculate the tieback load, we must first calculate the earth pressure acting on the walls, determine the tieback locations, and analyze the wall. The required capacity calculated in this way will be directly related to the assumptions made in selecting the earth pressure or analyzing the wall, which were discussed earlier. In general, the required capacity is calculated; then the tieback is designed to provide it, as discussed in earlier chapters.

The connection of the tieback to the wall must be designed. The pictures earlier in this chapter show several ways this is done. In general, this is a structural design detail and will not be discussed further, except to stress that it must be considered.

We do not want any portion of our anchors to be made in the soil between the wall and the critical failure surface. This is the soil, illustrated in Figure 10-1, which would move if the wall failed to support it. Clearly, any capacity the anchor developed in this soil would be reduced if the wall moved. We prevent this by specifying that the anchor be made some distance beyond the critical failure surface and designating the length of tieback which must be unbonded. We thus assure that the anchor is made beyond this soil. The physical significance of the unbonded length is to assure that no tendon load is transferred into this soil. This is the procedure I recommend.

However, some companies grout between the tendon and soil in the failure wedge, and some government agencies require such grouting. When bare bars are used with hollow-stem-auger tiebacks and the entire hole is filled with grout as the auger is withdrawn, the unbonded length of the tendon is zero. Then, when these tiebacks are tested, some of the measured anchor capacity is from the soil within the failure wedge. While we have not followed this design practice, I am intrigued with the fact that it has always been successful for others. Another practice, which has a similar result, is grouting the tendons in the unbonded length after they are tested and preloaded. Often this is accomplished by leaving a pipe around the tendon in this length and preventing grout from entering the pipe. Then, when the tieback is tested and preloaded, the pipe is filled with grout. This usually connects the soil to the tendon in this portion of its length. So far, I am not aware of any problem which has resulted directly from either practice of grouting the tendon to the soil between the wall and critical failure surface. Still, I would be uncomfortable doing it.

The tiebacks are anchored in a mass of soil, which must be stable. In general, the size of this soil mass is determined by the location of the tieback anchors. More confusion exists about this part of the design than any other, but the principles are reasonably simple.

The tiebacks tie the wall to a wedge of soil, which must be internally and externally stable. If the wall is properly designed and the tiebacks have the capacity for which they are designed, then the pressures on the wall and the tieback create internal forces in the wedge. The wedge must also be checked for external forces to verify its stability. The shape of this mass is fixed by the location of the tieback anchors, so we can assure a stable mass by properly locating the anchors. The object of Figure 10-14 is to illustrate that, in all practical cases, the tiebacks are an internal, unifying force in the wedge and we

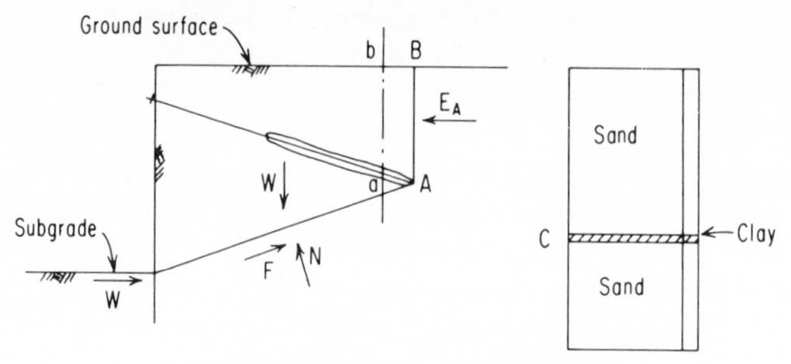

FIG. 10-14 Mass analyzed for stability.

must include all of the tieback anchor. The four sides of this wedge are the wall, the ground surface, a vertical plane through the ends of the tiebacks, and the weakest surface through the soil between the wall and the ends of the tiebacks.

Let's consider first the boundaries of the wedge. The wall and tiebacks act against the soil with forces that are equal and opposite, so no net force acts on this side of the wedge. The other side of the wedge is defined by the ends of the tiebacks. It should be clear that the wedge of soil will have no load acting on it from the tieback if this boundary goes through the end of the tieback, but, if it intersects the tieback, a stabilizing force will act. For example, on vertical plane *ab*, the earth pressure might be slightly less, but the tieback anchor is intersected and the force in the tieback at point *a* would act to resist sliding. Our experience has been that the least stable wedge is always the one where this edge is the plane between the end of the tieback and the surface. Tests done in Germany also show that the shape of the failure wedge is defined by the ends of the tiebacks. They caused a failure of model walls by pulling on the tiebacks. In all cases, when failure occurred, it passed through the ends of these tiebacks. If surcharges act on the ground surface, they are external forces. The fourth side of the wedge connects the ends of the tiebacks and the wall. This is the soil surface which must provide resistance to sliding, so we should use as our boundary the surface with the lowest net resistance to sliding. For example, this surface may be through a soft soil, such as the clay shown on the adjacent boring.

In general, the external force on this wedge, tending to displace it, is the earth pressure on plane *AB*, and the forces tending to resist displacing it act on the surface between *A* and the wall. The earth pressure on plane *AB* is usually taken as the computed active earth pressure, although some engineers feel the "at-rest" pressure should be used. In any case, the driving force is usually the earth pressure which acts on this plane. The bottom surface may not be straight, as shown in Figure

10-14, and must be fixed after examining the soil properties. For example, if a soft soil existed at elevation *C* on Figure 10-14, it would be reasonable to consider a failure through it. This must be checked carefully. In some cases, the tiebacks should penetrate below the soft soil to increase stability. The wall may also provide some resistance to sliding. In general, this is a fairly simple technique, which has yielded excellent results. In general, the surface with the lowest net resistance to sliding should be selected, and then the horizontal resistance to sliding calculated. To be stable, the resisting forces must exceed the driving forces by a suitable factor of safety.

From examination of this figure, it should be clear that the stability can be increased by increasing the length of tiebacks. This analysis is generally made to verify that the tieback length is adequate. If the mass is not stable, it will move and a crack may appear over the ends of the tiebacks. On projects in Seattle, Washington, D.C., and Atlanta, I have observed movement which I believe resulted from the tiebacks creating an internally stable, but externally unstable, mass. Increasing the number or capacity of tiebacks won't help these cases, but longer tiebacks will. This is important to understand: the tieback force is an internal force in this wedge and does not affect its external stability. We must design the wall and tiebacks to assure internal stability, which is what we did when we analyzed the earth pressure and designed the wall and tiebacks. Then we must be sure the mass of soil to which we have tied the wall is also stable.

The tieback applies a load on the wall, which pushes on the soil. In most cases, the tieback applies equal and opposite forces on the wall and anchor, which tend to squeeze the soil together. If we design for higher tieback capacities, both the force on the wall and the anchor are increased. The anchors are designed so they will not pull out of the soil, and so increasing the capacity of the tiebacks increases the squeezing of the wedge. Once enough capacity to assure stability is provided, additional capacity only serves to make a possible reduction in movement. These internal forces do not increase the external stability of the wedge.

We increase the wedge stability by modifying the tieback geometry. The driving forces can be reduced by making the end of the tieback shallower. The resisting forces can be increased by making the tieback longer or deeper or by making it penetrate through soft strata. Generally, we think in terms of longer tiebacks to accomplish this result.

Another procedure, frequently used in Germany, is to calculate the pull which would be required to dislodge the wedge of soil in which the tiebacks are anchored. This procedure is based on laboratory tests and analyses, in which the pull is applied to the wedge externally. This is seldom a realistic way to view the wedge, since both the wall and tiebacks support one side of the wedge and can have on them only the forces they apply to the earth wedge. They are always equal, except to the extent water pressure exists. Even then, the tiebacks must be designed to support the wall with water pressure on it, and this creates an internally stable wedge. I believe this is a procedure with limited application, which has been greatly overextended.

Tests to Verify Capacity of Individual Tiebacks

The testing program should consist of performance and proof tests, and should be used to assure each tieback will carry its design load. These tests, including testing for creep, were described in earlier chapters. It is important to realize that, if this testing program is followed, each tieback will be individually tested to an overload. Very few classes of structures are as carefully tested.

Tiebacks in some soils may creep. This will result in movement of the wall as the anchor moves or loss of tieback capacity if the tieback is overloaded. Field tests to analyze this problem were discussed earlier in this book. In general, all of these tests are based on using tiebacks only in soil where the rate of creep decreases with time and the magnitude of anchor creep over the lifetime of the structure is small.

Should this testing program identify an occasional tieback which will not carry its full design load, it is possible to use the test results to analyze that tieback further. Usually, it is reassigned a reduced load, and additional tiebacks are installed to provide the difference.

Occasionally the testing program leads us to conclude that the tieback design must be modified. Perhaps the tieback capacity selected was too high or low. The soil encountered on the project may be better or worse than the borings indicated. We have modified our tieback designs by changing tieback capacity, length, and angle to the wall, as well as the strata in which the tiebacks are made and even the method of making the tiebacks. The object of this testing program is to assure us that each tieback incorporated in a project will indeed carry the capacity assigned to it.

Tests to Verify Adequacy of Calculated Tieback Capacity

We have available simple tests to show that the tieback loads are not exceeding the design loads and measure the movement of the wall.

These were discussed in an earlier portion of this book. Load cells and strain gauges directly measure the tieback load. Surveys, extensometers, and slope indicators are used to measure wall movements. We can be quite confident of our design if the measured tieback loads do not exceed the design loads while wall movements are at an acceptable level.

Summary

Most tiebacks are used to support various types of walls. The walls are designed to provide a cover on the earth face and span between tiebacks. The tiebacks are designed to resist all or most of the earth pressure which acts on the wall. Each tieback is tested to verify its ability to carry the design loads. Simple tests are available to assure that the actual loads do not exceed the design loads and the wall does not move excessively.

11

LANDSLIDES, WALLS, AND TIEBACKS

Walls and tiebacks are being combined to stabilize landslides. These walls, in addition to supporting the soil behind them, must also apply a force on the sliding mass of soil which stabilizes it. Thus, the walls combine the functions described in the preceding chapter with the necessary strength to stabilize a landslide. The wall in Figure 11-1 must be stable when acted upon by the earth in the shaded wedge, but it must also apply a large enough force to stabilize the mass of soil above the landslide surface.

The wall is analyzed for two basic conditions. Primarily, it must apply a force to the large sliding mass which will stabilize it. This type of landslide involves a wedge of soil sliding on a definite failure surface. The wedge is stabilized by increasing the resisting forces which act on it. Tiebacks are a tool which can penetrate the surface of sliding, so they can exert the required force on the wedge. The anchors must be far enough beyond the sliding wedge that any other reasonable surface of sliding is also between them and the wall. Additional resistance to sliding can sometimes be achieved by driving or drilling wall elements across the slide surface.

The second condition for which the wall must be designed is as a retaining wall, as discussed in the preceding chapter. Obviously, in some cases the pressures on the wall from this condition may exceed those necessary to stabilize the landslide. It is important to recognize

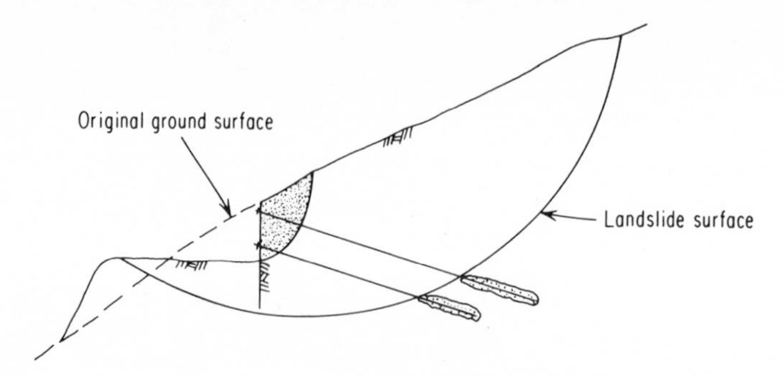

FIG. 11-1 Tied-back landslide-stabilization wall.

that the wall must be designed for both conditions, but in the rest of this chapter the discussion will be focused on the first condition only.

These walls provide an important new tool for landslide stabilization. The tiebacks provide a force which acts in a direction which is nearly ideal, and they can be installed without seriously aggravating the situation. Clearly, most of the tieback force in Figure 11-1 will act horizontally, but it has another component which will increase the normal force on the landslide surface. Both components act to stabilize the mass. Installation of the first row of tiebacks can even be coordinated with the excavation so that they act to provide a stabilizing force before any excavation takes place.

The three methods which have frequently been used to stabilize landslides in the past—flattening the upper slope, draining the groundwater to increase the soil strength on the landslide surface, or piling earth on the lower part of the slope to buttress it against sliding—are often neither as effective nor as economical as the wall. Flattening the upper slope often means a great deal of excavation, but has more serious drawbacks: it often cannot be done because the landslide is affecting an area of streets or buildings and the purpose of the corrective work is to protect these structures, not cause them to be removed. Increasing the soil strength on the surface of sliding by drainage is accomplished by reducing the buoyancy of the soil above it to increase its effective weight and thus the shearing resistance on the surface of sliding. This technique is often combined with installation of the wall to get the benefits of both. Buttressing the slide is also often impractical because it would interfere with the intended use or because a stream at the toe of the slide prevents building a buttress. In our experience, in comparison with conventional techniques, tiedback walls often have a clear advantage as the primary way to stabilize the landslide.

Classification of Landslides

Before proceeding to methods of analyzing landslides, we should consider classifying some types of landslides which can be stabilized with walls. Walls are most effective where the mass of soil which is sliding generally retains its structure, while sliding on a definite surface. Most of the methods available to analyze and correct landslides are based on the fact that this is the usual way they occur. One type of slide which has been well-documented involves a circular failure surface, which is usually in soft clay. More complex, noncircular, curved slide surfaces are also common. Another common slide involves the colluvium above the rock on mountainsides. Each of these common slide mechanisms will be discussed further. Another helpful distinction is between active and potential slides. The surface of sliding must be located as part of the analysis of the active landslide. Potential landslides are often caused by excavation or backfill, and in this case the analysis must determine the most probable sliding surface and the effect of the changes. Both of these conditions will be elaborated on. The soil always has some strength on the surface of sliding, and the design approach is to supplement this with the wall. As a result, we must first be able to analyze the forces causing and resisting sliding. Classifying the types of slides and their causes should be done as the first step.

The methods of evaluating landslides along curved surfaces have improved in recent years. Circular sliding masses are divided into segments, and the forces causing and resisting sliding analyzed, as shown in Figure 11-2. The center of rotation of the slide and the forces causing and resisting rotation are found. These techniques are explained in most soil mechanics textbooks and will not be fully discussed here. The salient fact is that techniques are available which should be used to analyze the landslide and can be used to evaluate the tieback forces required to increase stability to an acceptable level. With a circular failure surface, sliding of the mass can occur without internal distortions. In other words, as sliding occurs along a circular failure surface, the shape of the slide fits the shape of the sliding mass.

If the surface is a noncircular curve, distortion of the mass must occur as sliding continues. The best known method of analyzing a slide

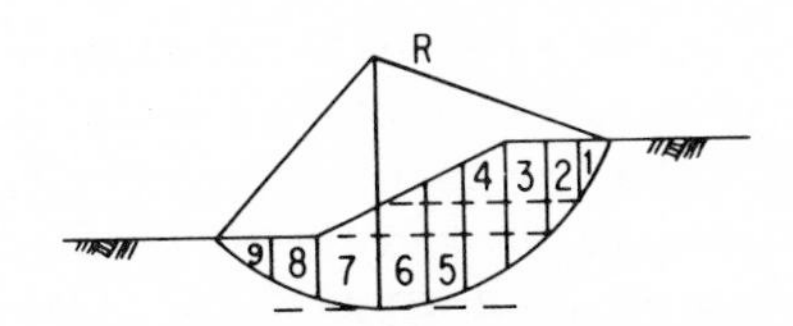

FIG. 11-2 Example analysis of circular landslide.

along a noncircular sliding surface is wedge analysis. The sliding mass is divided into a series of wedges; then the equilibrium of each wedge is calculated. Two more well-known solutions for this case are the Morgenstern-Price and Janbu procedures. Computers have made the use of these complex analytical techniques feasible. However, since walls are only now beginning to be used to stabilize landslides, practical examples of combining tiebacks with walls to prevent landslides are rare.

Colluvium slides on a mountainside usually occur where the soil lies on the rock surface. Frequently, water that has seeped into the soil weakens the earth just above the rock. The soil overlying the rock is often of a more or less uniform thickness, so that this type of slide can often extend a long way up the mountain. The surface of sliding appears more like a smooth plane than a curved surface. This is the second common type of landslide which can be stabilized with a wall. Next, let's distinguish between active and induced landslides.

Most active landslides are moving at a slow rate on a definite failure surface. Often, movement occurs only during wet seasons or periods of heavy rain. Obviously, walls can only be built to stabilize slow-moving slides, since their construction may require months of work. Surveys can establish the rate of movement. Slope indicators are one of the most effective ways of locating the surface of sliding and establishing the rate of movement there. Usually an experienced engineer can identify the limits of the slide area by finding a scarp at the top of the slide and a berm at the toe. Slope indicators and borings can be combined to locate the sliding surface between these points. These field observations can be combined to establish the shape of the sliding mass, the surface on which it is sliding, and the rate of sliding. This information must be available for analyzing the slide and planning the corrective work.

Some walls are designed to prevent landslides from occurring as a result of construction which might trigger the slide. In this case, the surface on which sliding is most likely to occur must first be identified. Since a slide surface does not exist, various surfaces are investigated until the critical one is found. At different stages in the construction there may be different surfaces, so that the stability of the soil must be checked at each critical stage. Clearly, excavating soil at the toe of a slope may cause this type of slide by removing soil which was necessary for the stability of the slope. Adding weight to the slope, for example by backfilling near the top, may cause it to slide. In stiff clays, a gradual swelling of the soil caused by removing soil which confined it, may trigger a slide. In general, this type of slide must be anticipated in the design phase of the project and properly designed for.

The first step in designing a wall to correct a landslide is to properly classify the landslide. Walls have been most effective where there is a definite surface on which the landslide occurs, so that a force applied to the wall stabilizes the sliding wedge of soil.

Landslide Analysis

The next step in designing the corrective wall is to analyze the landslide, so that the deficiency which the wall must correct for can be established. The deficiency will be corrected by the tiebacks, which pass through the surface of sliding, or the wall. In the preceding discussion, landslides were classified in several ways, and the analysis for each classification is different. Other classifications might also affect the method of analysis, but only four classifications will be discussed because they are so common.

When analyzing a circular failure surface, the radius of curvature is first established. Then the mass of soil is analyzed to calculate the moment at the failure surface which tends to drive the landslide. The moment of the shear force on this surface is calculated, as well as any other moment tending to resist the slide. These techniques are explained in detail in most soil mechanics texts, so they will not be fully discussed here. The results of this analysis will be a radius of the circular failure surface and moments acting along that surface causing and resisting the slide.

Noncircular curved sliding surfaces are more difficult to analyze, but techniques have been developed which facilitate such analysis. For example, the techniques of Janbu, Morgenstern-Price, and others can be used. Some of these techniques have been computerized to simplify the extensive calculations involved. The details of these methods are reviewed in other publications, so no attempt will be made here to explain them. The important point is to know that this analysis also results in a comparison of driving and resisting forces and these must be computed before the walls can be analyzed.

The other shape of sliding surface discussed was a plane. In this case, the slide is divided into slices, and one slice is analyzed. A typical slice is shown in Figure 11-3. The slice is then analyzed in a similar manner to determine driving and resisting forces on the surface of sliding. When these are known for the slice, the length of the slide is next established, so that the total sliding and resisting forces can be calculated. In the same way as the earlier cases, these values must be known before the wall strength can be calculated.

In each of these cases, both the surface of sliding and the strength of

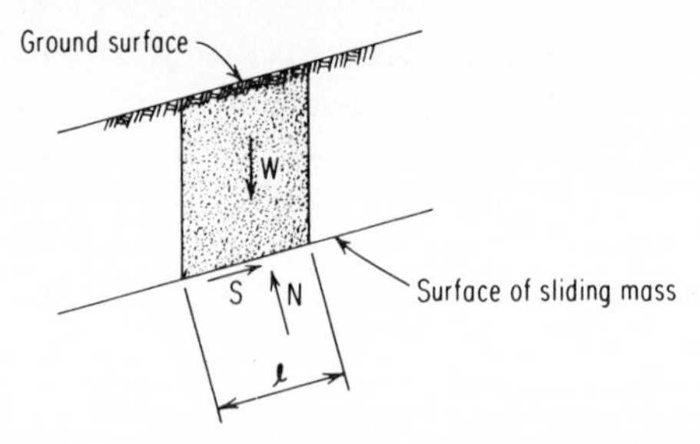

FIG. 11-3 Forces acting on idealized slide on infinite slope.

the soil on that surface must be established. A great deal of soil mechanics literature deals with selecting the appropriate laboratory tests for determining these properties. Where a slide has not yet occurred, these tests generally provide the only estimate of soil strength available.

When a slide has already occurred, it is customary also to calculate the soil properties for a factor of safety of 1. Since both cohesion and friction resist the slide, it is often difficult to separate their contributions. Often when engineers analyze the resisting force, they assume it to be totally due to first cohesion and then friction and thus establish upper bounds for each of these properties. Of course to do this, they must first establish the surface of sliding and extent of the slide. In this way, the actual slide can be utilized to obtain information on the strength of the soils on the sliding surface.

One point that must be considered is that the strength on the failure surface should not decrease after the wall is built. The basic assumption made in the design of these walls is that the wall will supplement the soil strength on the sliding surface. If movement of the slide continues, the force in the tiebacks will increase and possibly lead to wall failure. The landslide must be stabilized by the wall, and in general the resisting forces divided between the soil and wall must be stable. The analysis should address this point and verify that the soil strength along the surface of sliding will not decrease with time.

Wall and Tieback Design

In the preceding analysis, the landslide has been analyzed for its driving and resisting forces. The wall and tiebacks are then designed to provide an additional desirable amount of resistance.

It is usually possible to accomplish this with a wall at many different locations. The engineer should first establish a reasonable wall location to solve the problem of the landslide. In some cases, this may be a wall location near the top of the landslide, with regrading of the slope below to flatten it. In other cases, it may be fixed by the alignment of a road or railroad cut causing the slide. Several wall locations may be combined effectively in other cases. The first step in the analysis of the wall and tiebacks is to locate the wall, usually so that the tiebacks will be most effective.

Some walls penetrate the surface of sliding, and some don't. If the wall does penetrate, it can often be designed for support in the soil below the slide with only one tieback at the top. The wall is then designed to provide some shear across the sliding surface. The tieback at the top provides additional force on the mass to stabilize it. The wall may be designed to span between these two supports. If the depth of the slide is shallow, this may be very satisfactory. For deep slides, such a wall must be extremely rigid and expensive to be effective. When the wall does not penetrate, the tiebacks can be used both to support the wall and to stabilize the landslide. In this case, the entire stabilizing force must be provided by the tiebacks. The configuration of the tiebacks should be studied to provide the greatest economy. In some cases a continuous wall is not required, and tiebacks through separate blocks are adequate. In most cases, at least two rows of tiebacks will be placed against the wall to assure it is stable. When this is done, the magnitude of the tieback force is calculated, and then compared with the magnitude of the force calculated using the methods of the preceding chapter. The wall is designed for the larger force. If the forces in the tiebacks to stabilize the landslide are larger, then this is distributed as a uniform pressure on the wall. The wall is sized for this applied pressure.

The forces causing sliding in Figure 11-4 are analyzed, and the horizontal reaction is summarized as D. The forces resisting sliding are similarly analyzed, and the horizontal resistance summarized as R.

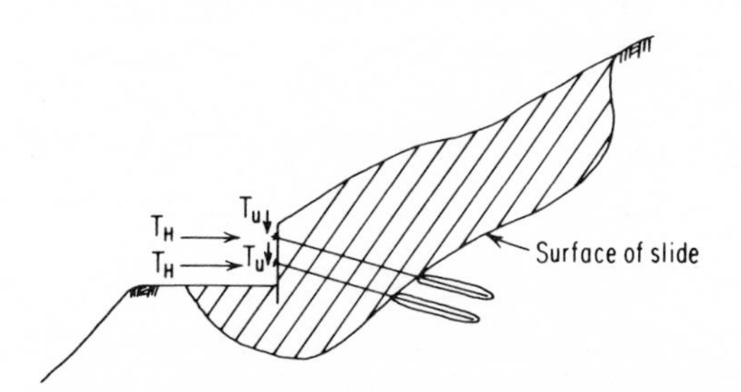

FIG. 11-4 Tieback force required to stabilize landslide.

The horizontal reactions of the tiebacks to provide the desired supplemental horizontal force are determined. The vertical component will, of course, act to increase the normal force on the sliding surface, and may be considered, therefore, to increase the shear on this surface. The factor of safety against sliding will then be:

$$F = \frac{\Sigma R + \Sigma T_H}{\Sigma D}$$

This is the general approach to determining the magnitude of the tieback force required. Of course, good design involves the best use of the tiebacks to stabilize the landslide. This will consider the interrelationship of angle, length, and capacity. Clearly, if the analysis is based on comparing horizontal forces, tiebacks will be most effective if they can be nearly horizontal. The total tieback capacity will not have to be as high as with steeper tiebacks. In general, at an angle of about 15°, 95 percent of the tieback load will act horizontally. Flat tiebacks will generally be longer than steeper ones, as can be seen by observing Figure 11-4. This will usually increase their cost. So a balance between length, angle, and capacity should be achieved.

Some engineers have used tiebacks in a completely opposite approach. They are used at a steep angle to cross the sliding surface at right angles. Then the tieback load increases the sliding resistance by increasing the normal forces. Or they are used near the toe of the slope to replace the buttressing. In these cases, the tiebacks really act not against a wall, but instead on the surface of the mass.

The tieback ties the sliding mass of soil to the earth beneath and thus creates a larger mass of soil, which must also be checked for stability. In some cases, as when the underlying soil is rock, this may be no problem. In other cases, this enlarged mass must be analyzed for stability. In Figure 11-5, the most probable potential sliding surface through the ends of the tiebacks is shown. The factor of safety for this condition must equal or exceed the factor of safety with the wall on the original surface. We can assure this by making the tiebacks long

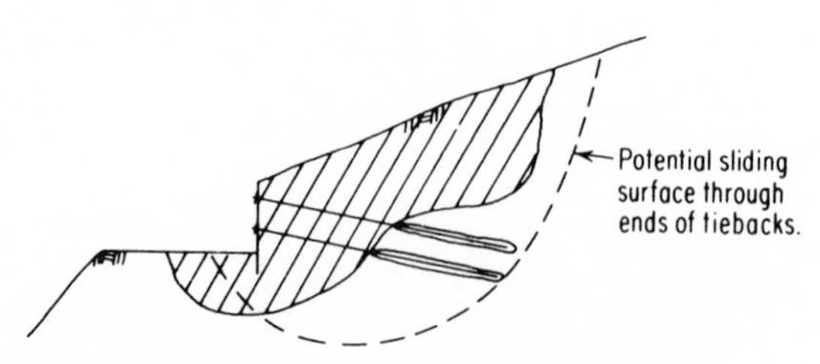

FIG. 11-5 Stability analysis of tied-back wall.

enough. The final step in the design is to make sure the anchors are deep enough in the stable soil.

When tiebacks are used to stabilize a landslide, there may be no reason for an unbonded length. Since the failure surface is just that, it may even be desirable to bond the tiebacks to the soil on both sides of the sliding surface. If this is done, then the tendons and grout not only provide keys across the surface of sliding, but even a slight movement will mobilize the full tension on the tieback. Perhaps preloading is not needed in this case. When the sliding mass is large and coherent, it may not even be necessary to provide a wall.

Summary

Before tiebacks, walls were not usually effective in stabilizing landslides, but now tiebacks provide a powerful new tool. The tiebacks go through the sliding mass and surface of sliding to tie the sliding mass to the stable soil below. We can design the tiebacks to provide the stabilizing force and locate them to provide it at the most desirable place. We can also, if we choose, use them to increase the soil strength. The tiebacks act on the sliding mass of soil to stabilize it, but are anchored in stable soil.

Conclusion

My goal in this book has been a realistic presentation of our present state of knowledge about tiebacks and their use, based both on my awareness of what others are doing in the field and on my own years of successful experience. Years ago, a fine construction man said that education was learning from the experience of others. It is my hope that I have succeeded in making you more comfortable with the design, construction, and use of tiebacks and that you will share some of my enthusiasm for this new foundation engineering tool.

BIBLIOGRAPHY

The bibliography which follows has been included to help the reader who wishes additional information on tiebacks. It is not complete, but includes articles which have been quoted in the book, as well as articles which present contrary views. Several of the references have extensive bibliographies which list additional sources of information on tiebacks.

Bastable, A. D. (1974)
"Multibell Ground Anchors in London Clay," a paper presented at the 7th Congress of the Fédération Internationale de la Précontrainte, Prestressed Foundation and Ground Anchors, New York, 1974, pp. 33–38.

Bauer, K.
The Bauer Injection Anchor and Its Applications, Associated Tunnelling Co., Ltd., Lowton St. Mary's, Near Warrington, Lancs., England.

Bethlehem Steel (1977)
How Corrosion Affects Steel H-Piles Buried in Earth, Bethlehem, Pa., 1977.

Bjerrum, L., C. J. F. Clausen, and J. M. Duncan (1972)
"Earth Pressures on Flexible Structures: A State-of-the-Art Report," *Proceedings of the 5th European Conference on Soil Mechanics and Foundation Engineering*, vol. 2, Madrid, 1972.

Bureau Securitas (1977)
Recommendations on the Design, Calculation, Construction, and Control of Tiebacks (Recommendations concernant le conception, le calcul, l'execution et la controle des tirants d'ancrage), Paris, 1977.

Chapman, K. R. (1975)
"Specifications for Earth Tieback Sheeting and Tieback Testing Procedures," *The Construction Specifier,* June 1975.

Clifton, J. R., H. F. Beeghly, and R. G. Mathley (1975)
Non-metallic Coatings for Concrete Reinforcing Bars, Building Science Series 65, U.S. Department of Commerce N.B.S., 1975.

Clough, G. W. (1974)
"Tied-Back Walls," a workshop session report, *Proceedings of the Conference on Analysis and Design in Geotechnical Engineering,* vol. 2, University of Texas, Austin, June 1974, pp. 100–102.

"Controlling Landslides," *Compressed Air Magazine*, November 1977, pp. 16–17.

Costa Nunes, A. J. (1969)
"Anchorage Tests in Clays for the Construction of Sao Paulo Subway," Paper 15-2, *7th International Conference of the International Society on Soil Mechanics and Foundation Engineering,* Mexico, 1969, pp. 120–125.

German Codes (Deutsche Normen), DIN 4125 (1972)
Grouted Temporary Soil Tiebacks, (*Verpressanker für vorübergehende Zwecke in Lockergerstein*), Beuth-Vertrieb GmbH, Berlin, 1972.

Dupeuble, P., and J. Brulois (1969)
"Measurement and Observation on the Behavior of Anchorages in Soils," Special Sessions Nos. 14–15, International Society on Soil Mechanics, Mexico, 1969, pp. 147–155.

Gandias, M. (1977)
"Permanent Ground Anchors and Ground Anchor Tests" (Tirant definitifs et essais de tirants), *Annales de l'institut technique du batiment et des travaux publics*, 346, January 1977.

Goldberg, D. T., W. E. Jaworski, and M. D. Gordon (1976)
Lateral Support Systems and Underpinning (three volumes), Federal Highway Administration Reports, 1976.
FHWA-RD-75-128 (Vol. I, *Design & Construction*)
FHWA-RD-75-129 (Vol. II, *Design Fundamentals*)
FHWA-RD-75-130 (Vol. III, *Construction Methods*)

Griess, J. C., and D. J. Naus (1975)
"Corrosion of Steel Tendons Used in Prestressed Concrete Pressure Vessels," National Technical Information Service Report on research sponsored jointly by U.S. Nuclear Regulatory Commission and U.S. Department of Energy, 1975.

"Ground Anchors" (1974)
Commission Reports, *Proceedings of the 7th Congress of the Fédération Internationale de la Précontrainte,* New York, 1974.

Hanna, T. H. (1976)
"Analysis Design and Installation of Tiebacks and Ground Anchors," *Proceedings of the Seminar on Underground Construction Problems Techniques and Solutions*, Chicago, October 20–22, 1975, pp. 306–319.

Herbst, T. (1978)
"On the Technology of Load Transfer Between High Strength Prestressing Threadbars and the Ground as Experienced with Monobar Soil Anchors," presented at 6th Brazilian Conference on Soil Mechanics and Foundation Engineering, Rio de Janeiro, 1978.

Hilton, A. R. (1969)
"The Construction of High Strength Under-reamed Anchorages in Soft Ground," Special Sessions Nos. 14–15, 7th International Conference of Society on Soil Mechanics and Foundation Engineering, Mexico, 1969, pp. 172–174.

Jelinek, R. and H. Ostemayer (1976)
"Grouted Anchors in Soil" (Verpressanker in Boden), *Bauingenicor*, vol. 51, no. 2, 1976, pp. 109–118.

Khaova, M., B. Montel, A. Civard, and R. Lauga (1969)
"Cheurfas Dam Anchorage: 30 Years of Controls and Recent Reinforcement," Proceedings of the 7th International Conference on Soil Mechanics and Foundation Engineering, Paper 15–12, 1969, pp. 167–171.

Kramer, H. (1977)
"Determination of the Carrying Capacity of Ground Anchors with the Correlation and Regression Analysis," Special Session No. 4, *Ground Anchors*, 9th International Conference on Soil Mechanics and Foundation Engineering, Tokyo, July 1977.

Littlejohn, G. S., and D. A. Bruce (1977)
Rock Anchors: State-of-the-Art, Foundation Publication Ltd., Essex, England, 1977.

McClenahan, M. R. C. (1976)
"Ground Anchors at Lowestoff," *Ground Engineering*, November 1976.

Ostermayer, H. (1975)
"Construction Carrying Behavior and Creep Characteristics of Ground Anchors," Conference: Diaphragm Walls and Anchorages, Institution of Civil Engineers, London, 1975, pp. 141–151.

Prestressed Concrete Institute (1974)
"Tentative Recommendations for Prestressed Rock and Soil Anchors," ad hoc committee of the PCI Post-tensioning Committee, Prestressed Concrete Institute, Chicago, March 1974.

"Permanent Tiebacks Winning an Uphill Battle" (1978)
ENR, October 26, 1978, p. 19.

Portier, J. L. (1974)
"Protection of Tiebacks Against Corrosion," paper presented at the 7th Congress of the Fédération Internationale de la Précontrainte, Prestressed Concrete Foundations and Ground Anchors, New York 1974, pp. 39–53.

Ranke, A. and H. Ostermayer (1968)
"A Contribution to the Stability Calculations of Multiple Tied-Back Walls" ("Beitrag zur Stabilitäts Untersuchung . . ."), *Die Bautechnik*, vol. 45, no. 10, 1968, pp. 341–349.

Reese, L. C., F. T. Touma, and M. W. O'Neill (1976)
"Behavior of Drilled Piers under Axial Loading," *ASCE Journal of the Geotechnical Engineering Division*, GT5, May 1976, pp. 493–510.

Rehm, G. (1971)
"Corrosion Protection for Injection Anchors," *Vortäge Baugrundtagung der Deutschen Gesselschaft für Erd- und Grundbau,* Essen, 1971, pp. 37–35.

Romanoff, M. (1957)
"Underground Corrosion," National Bureau of Standards Circular 579, NTIS PB168350, April 1957.

Schnabel, Jr., H. (1971)
"Sloped Sheeting," *Civil Engineering ASCE*, February 1971, pp. 48–51.

Schupack, M. (1978)
A Survey of the Durability Performance of Post-Tensioning Tendons, Post Tensioning Institute, January 1978.

Shields, D. R., H. Schnabel, and D. E. Weatherby (1978)
"Load Transfer in Pressure Injected Anchors," Geotechnical Engineering Division, *Proceedings ASCE* 104, no. GT9, 1978, pp. 1183–1196.

Sigourney, J. W. (1971)
"Tieback Installations," paper presented at Conference on Design and Construction of Earth and Rock Tiebacks, George Washington University, Washington, D.C., February 3, 1971.

Terzaghi, K., and R. B. Peck (1948)
Soil Mechanics in Engineering Practice, John Wiley, New York, 1948.

Weatherby, D. E. (1980)
"Tiebacks and Their Applications," talk given at *Geotech '80*, Chicago, March 17, 1980.

——— "Soil and Rock Anchor Tiebacks," talk given at Construction Excavation Technical Institute, University of Wisconsin, September 8–9, 1980.

INDEX